AF876727

Springer Complexity

Springer Complexity is an interdisciplinary program publishing the best research and academic-level teaching on both fundamental and applied aspects of complex systems—cutting across all traditional disciplines of the natural and life sciences, engineering, economics, medicine, neuroscience, social and computer science.

Complex Systems are systems that comprise many interacting parts with the ability to generate a new quality of macroscopic collective behavior the manifestations of which are the spontaneous formation of distinctive temporal, spatial or functional structures. Models of such systems can be successfully mapped onto quite diverse "real-life" situations like the climate, the coherent emission of light from lasers, chemical reaction-diffusion systems, biological cellular networks, the dynamics of stock markets and of the internet, earthquake statistics and prediction, freeway traffic, the human brain, or the formation of opinions in social systems, to name just some of the popular applications.

Although their scope and methodologies overlap somewhat, one can distinguish the following main concepts and tools: self-organization, nonlinear dynamics, synergetics, turbulence, dynamical systems, catastrophes, instabilities, stochastic processes, chaos, graphs and networks, cellular automata, adaptive systems, genetic algorithms and computational intelligence.

The three major book publication platforms of the Springer Complexity program are the monograph series "Understanding Complex Systems" focusing on the various applications of complexity, the "Springer Series in Synergetics", which is devoted to the quantitative theoretical and methodological foundations, and the "Springer Briefs in Complexity" which are concise and topical working reports, case studies, surveys, essays and lecture notes of relevance to the field. In addition to the books in these two core series, the program also incorporates individual titles ranging from textbooks to major reference works.

Indexed by SCOPUS, INSPEC, zbMATH, SCImago.

Series Editors

Henry D. I. Abarbanel, Institute for Nonlinear Science, University of California, San Diego, La Jolla, CA, USA
Dan Braha, New England Complex Systems Institute, University of Massachusetts, Dartmouth, USA
Péter Érdi, Center for Complex Systems Studies, Kalamazoo College, Kalamazoo, USA;
Hungarian Academy of Sciences, Budapest, Hungary
Karl J. Friston, Institute of Cognitive Neuroscience, University College London, London, UK
Sten Grillner, Department of Neuroscience, Karolinska Institutet, Stockholm, Sweden
Hermann Haken, Center of Synergetics, University of Stuttgart, Stuttgart, Germany
Viktor Jirsa, Centre National de la Recherche Scientifique (CNRS), Université de la Méditerranée, Marseille, France
Janusz Kacprzyk, Systems Research Institute, Polish Academy of Sciences, Warsaw, Poland
Kunihiko Kaneko, Research Center for Complex Systems Biology, The University of Tokyo, Tokyo, Japan
Markus Kirkilionis, Mathematics Institute and Centre for Complex Systems, University of Warwick, Coventry, UK
Ronaldo Menezes, Department of Computer Science, University of Exeter, UK
Jürgen Kurths, Nonlinear Dynamics Group, University of Potsdam, Potsdam, Germany
Andrzej Nowak, Department of Psychology, Warsaw University, Warszawa, Poland
Hassan Qudrat-Ullah, School of Administrative Studies, York University, Toronto, Canada
Linda Reichl, Center for Complex Quantum Systems, University of Texas, Austin, USA
Peter Schuster, Theoretical Chemistry and Structural Biology, University of Vienna, Vienna, Austria
Frank Schweitzer, System Design, ETH Zürich, Zürich, Switzerland
Didier Sornette, Institute of Risk Analysis, Prediction and Management, Southern University of Science and Technology, Shenzhen, China
Stefan Thurner, Section for Science of Complex Systems, Medical University of Vienna, Vienna, Austria

Understanding Complex Systems

Founding Editor: Scott Kelso

Kristian Lindgren

Information Theory for Complex Systems

An Information Perspective on Complexity in Dynamical Systems and Statistical Mechanics

Springer

Kristian Lindgren
Department of Space, Earth
and Environment
Chalmers University of Technology
Gothenburg, Sweden

ISSN 1860-0832 ISSN 1860-0840 (electronic)
Understanding Complex Systems
ISBN 978-3-662-68212-8 ISBN 978-3-662-68214-2 (eBook)
https://doi.org/10.1007/978-3-662-68214-2

© Springer-Verlag GmbH Germany 2024

This work is subject to copyright. All rights are solely and exclusively licensed by the Publisher, whether the whole or part of the material is concerned, specifically the rights of translation, reprinting, reuse of illustrations, recitation, broadcasting, reproduction on microfilms or in any other physical way, and transmission or information storage and retrieval, electronic adaptation, computer software, or by similar or dissimilar methodology now known or hereafter developed.
The use of general descriptive names, registered names, trademarks, service marks, etc. in this publication does not imply, even in the absence of a specific statement, that such names are exempt from the relevant protective laws and regulations and therefore free for general use.
The publisher, the authors, and the editors are safe to assume that the advice and information in this book are believed to be true and accurate at the date of publication. Neither the publisher nor the authors or the editors give a warranty, expressed or implied, with respect to the material contained herein or for any errors or omissions that may have been made. The publisher remains neutral with regard to jurisdictional claims in published maps and institutional affiliations.

This Springer imprint is published by the registered company Springer-Verlag GmbH, DE, part of Springer Nature.
The registered company address is: Heidelberger Platz 3, 14197 Berlin, Germany

Paper in this product is recyclable.

To Eva,
for love and encouragement,
making everything possible.

Preface

This book is the result of a forty-year journey of collaboration and thoughtful discussions enriched by insights from colleagues, teachers, and students.

A key person and a source of inspiration is Karl-Erik Eriksson who introduced me to physics and information theory in 1983 as supervisor for my M.Sc. and Ph.D. theses. Our collaboration lead to the information-theoretical framework that serves as a basis for this book. The subsequent expansion and refinement of the theory, as well as its diverse applications, were achieved through numerous research projects. I am grateful for the fruitful and enjoyable collaborations with Karl-Erik Eriksson, Bengt Månsson, Mats Nordahl, Anders Eriksson, Torbjørn Helvik, and Eckehard Olbrich.

The stimulating research environment at the Division of Physical Resource Theory, including the Complex Systems Group, at Chalmers University of Technology, and the experiences gained during visits to several research institutes have been instrumental in refining the material presented in this book. I extend my gratitude for the welcoming atmospheres at the Santa Fe Institute, the Max Planck Institute for Mathematics in the Sciences, the Nordic Institute for Theoretical Physics (NORDITA), and the European Centre for Living Technology. The invaluable input and critical discussions from numerous researchers, including Nihat Ay, Sven Banisch, Nils Bertschinger, Jim Crutchfield, Doyne Farmer, David Feldman, Martin Nilsson Jacobi, Jürgen Jost, Sten Karlsson, Tomas Kåberger, John McCaskill, Cris Moore, Norman Packard, Steen Rasmussen, and Cosma Shalizi, have been critical at various stages of this work.

Since 1990, I have taught a graduate course, titled 'Information Theory for Complex Systems,' at Chalmers. Over the past three decades, the lecture notes for this course have matured alongside the research, resulting in the creation of this book. Starting in 2000, when the Complex Adaptive Systems master's programme began at Chalmers, this course has provided a theoretical basis for the study of complex systems. Each chapter of the book includes a set of problems of varying difficulty levels, with many drawn from past exam problems in the course.

Several Ph.D. and graduate students have contributed to the book's development, some of whom have also served as teachers for problem sessions. Notably, I want to mention Erik Edlund, Rasmus Einarsson, Olof Görnerup, Oskar Lindgren, Johan Lindgren, Susanne Pettersson, Ingrid Råde, and Ankit Vikrant, who has graciously taken over the role as the main teacher for the course.

Flatön, Sweden
July 2023

Kristian Lindgren

Contents

Chapter 1
Introduction

The term "complex systems" has become a unifying concept for research and studies of large multi-component systems in many disciplines such as physics, chemistry, biology, social science, computer science, economics, and geography. The interdisciplinary nature of these studies has resulted in a wide range of applications of complex systems. In physics, examples range from the second law of thermodynamics as a statistical result of large physical systems to self-organizing systems in the form of chemical reaction-diffusion systems.

Cellular automata provides an illustrative example of an abstract class of computational systems that exhibit various complex phenomena. Additionally, biologically inspired algorithms, like those found in artificial neural networks and evolutionary computation, offer powerful tools for learning and optimization.

The study of collective phenomena in societal systems, such as economic interactions among agents, the evolution of cooperation, urban growth, and traffic systems, has also been a subject of great interest within the realm of complex systems.

Over the past four decades, the area of complex systems has grown tremendously, leading to a large number of scientific journals. This growth can be attributed to several factors, including the rapid development of computers, which has allowed for cheap and powerful experimental laboratories for complex systems modelling. Furthermore, the study of complex systems has led to important discoveries and insights into the behaviour of various natural and artificial systems, and has become an essential field of study in many areas of research.

A complex system typically consists of a large number of components, which may be simple in terms of their internal characteristics and their interaction with each other. However, when the system is observed over extended time and length scales, there may be emergent phenomena that are not easily explained by the individual components and their interactions. These phenomena may present new challenges that require a deeper understanding of the system's behaviour.

There is no universal definition of a complex system, but researchers generally consider several features when describing a system as complex. One important

© Springer-Verlag GmbH Germany 2024

K. Lindgren, *Information Theory for Complex Systems*, Understanding Complex Systems, https://doi.org/10.1007/978-3-662-68214-2_1

scientific question is whether various characteristics of complex systems can be quantified. The aim of this book is to provide a set of tools that can be used to give quantitative characterisations of complex systems in different situations.

Statistical methods are important tools for analysing systems composed of many components. Within physics, this approach is taken in statistical mechanics. Information theory, a branch of probability theory, has a mathematical basis equivalent to a part of the foundations of statistical mechanics (Jaynes 1957). Information theory also provides concepts and methods that are useful in analysing structure, disorder, randomness, and other aspects of models of complex systems.

Information-theoretic concepts can be applied on the macro-level of a system, such as to describe the spatial structure formed in a chemical self-organising system. The connection between information theory and statistical mechanics allows such an analysis to be related to the thermodynamic properties and limitations of the system.

Information theory may also be applied on the micro-level in physical systems, such as in spin system models and other more abstract models for statistical mechanics. In this book, it is illustrated how information-theoretic quantities can be related to statistical mechanics and thermodynamic properties. In this way, we can make information-theoretic illustrations and explanations of several complex phenomena like the second law of thermodynamics.

The development of information quantities in the time evolution of various systems is a key focus in this book. A fundamental question for the time evolution is whether the dynamics can be considered deterministic and reversible. In many models of physical systems one uses probabilistic equations for the time development. This has direct implications for various information characteristics, since the probabilistic component can be viewed as an inflow of (random) information. One should note though, that all fundamental equations of motion in physical systems are deterministic and reversible. How can this be consistent with the second law of thermodynamics stating that disorder (entropy) is increasing for a closed system? This and related questions will be investigated in detail throughout the book.

The concept of information has multiple meanings. For a historical survey and a discussion of its usage in different scientific disciplines, see the work by Capurro and Hjørland (2003). During the 20th century, information has often been associated with knowledge or the transmission of knowledge. The introduction of information as a quantity in science and engineering was linked to the development of communication technology. In the 1920s, Harry Nyquist and Ralph Hartley pioneered this field and published papers in the context of telegraph speed and "information" transmission, providing the first quantitative definition of information (Nyquist 1924; Hartley 1928). Their ideas were based on the observation that this quantitative measure should be proportional to the logarithm of the number of choices or possibilities for composing the message.

This basic quantity was later generalised by Claude Shannon and Warren Weaver when they presented the information theory that serves as the foundation of the field today (Shannon and Weaver 1949). This generalisation is based on a probabilistic or statistical description of the system under study.

It is important to note that the concept of information presented in this book is a specific one, related to a probabilistic description of the studied system or how such a description changes when the knowledge about the system is updated. This means that what is quantified is related to the description of the system, rather than the system itself. One can choose a description that hides many details and, in that case, may get a different information quantity compared to the fully described system.

It is worth noting that the information we consider has no direct connection to meaning, as stated in Shannon's first paper presenting the theory (Shannon 1948):

"... the fundamental problem of communication is that of reproducing at one point either exactly or approximately a message selected at another point. Frequently the messages have meaning; that is they refer to or are correlated according to some system with certain physical or conceptual entities. These semantic aspects of communication are irrelevant to the engineering problem. The significant aspect is that the actual message is one selected from a set of possible messages."

The information-theoretic basis developed in this book was first presented in a series of papers and a book in the 1980s (Eriksson 1985; Eriksson et al. 1987; Eriksson and Lindgren 1987; Lindgren 1987; Lindgren and Nordahl 1988; Lindgren 1988, 1989). Further development of the theory and some applications have been made in subsequent years (Lindgren et al. 2004; Helvik et al. 2007; Helvik and Lindgren 2014; Lindgren 2015; Lindgren and Olbrich 2017). Originally, the application areas considered were broader, but here we focus on a few complex systems related areas: cellular automata, statistical mechanics and thermodynamics, chemical pattern formation, and chaotic systems.

The book is organised as follows. In Chap. 2, basic concepts in information theory are presented along with the maximum entropy principle. Chapter 3 presents the application of the concepts to symbol sequences or, more generally, to lattice systems in one or more dimensions. Complexity measures formulated in information-theoretic terms are presented. The formalism is applied to simple models of stochastic symbol sequence generation.

In Chap. 4, the information-theoretic framework is applied to the class of discrete dynamical systems called Cellular Automata (CA). Information theory provides several tools for studying the dynamics and patterns generated. We will primarily focus on one-dimensional CA for the analysis of symbol sequences that represent states in a CA. The aim is to provide explanations for how order and disorder may develop in the time evolution. In Chap. 5, some connections between the information-theoretic quantities and physics (statistical mechanics and thermodynamics) are presented. This relation makes it possible to connect some of the information-theoretic quantities describing a system with the thermodynamic laws that put constraints on the dynamics of the system. The analysis of a dynamical spin system model provides an explanation for how disorder may seemingly increase despite the fact that the underlying dynamics is fully reversible.

Chapter 6 introduces geometric information theory, which enables analysis of spatially extended patterns in continuous space. This approach can be applied to identify the location of information in images, both in terms of position and length scale. The chapter also includes an information-theoretic characterisation of fractal patterns and their dimension properties. In Chap. 7, we apply geometric information theory to pattern formation in chemical systems. The physical constraints of a chemical self-organizing system is translated to corresponding information-theoretic properties of the system and its dynamics. We derive a continuity equation for information that describes the flow of free energy driving the pattern formation process. In Chap. 8, we explore the information-theoretic perspective on chaotic dynamical systems, showing how sensitivity to small fluctuations can be seen as an information flow from "micro" to "macro". Here we use information theory for symbol sequences in order to analyse various examples of simple chaotic systems.

Chapter 2
Information Theory

Abstract The basic concepts of information theory are presented. These are all based on a probabilistic description of the system that is observed and characterised. The quantities *information*, *entropy*, and *relative information* are the key concepts that will be used throughout the book. The maximum entropy formalism is introduced, which provides an unbiased probabilistic description of a system, given some knowledge about it. Several examples and problems are provided to help explore and understand the basic concepts.

2.1 Basic Concepts

The notion of a quantitative measure of information was initially proposed in the 1920s by Harry Nyquist (1924) and Ralph Hartley (1928), both papers published in the Bell System Technical Journal. Hartley showed, in his paper on signal transmission, that the information content I_H in a message consisting of n characters, each of them chosen from an alphabet of N different symbols, should be proportional to the length n multiplied by some function of the size of the alphabet, $f(N)$. Hence, $I_H = nf(N)$. The functional form can be derived from the argument that if the message is coded using an alphabet of different size N', resulting in different length n', the information content should remain consistent, i.e., $I_H = n'f(N')$.

Consider an example where the new alphabet has $N' = N^2$. In this case, the length is reduced to $n' = n/2$, which is achieved by substituting pairs of symbols from the old alphabet with single symbols from the new one. Thus, the information quantity becomes $I_H = n/2 f(N^2) = nf(N)$. The function f that ensures preservation of the information content is $\log_K$, where K is an arbitrary base of the logarithm. Changing K merely scales the information content. Typically, we use base 2 logarithm, denoted as log, which expresses information quantity in binary units or *bits*,

$$I_H(n, N) = n \log N \; . \tag{2.1}$$

© Springer-Verlag GmbH Germany 2024
K. Lindgren, *Information Theory for Complex Systems*, Understanding Complex Systems, https://doi.org/10.1007/978-3-662-68214-2_2

The information per symbol is then

$$I_H(N) = \log N \ . \tag{2.2}$$

Hence, a binary alphabet (e.g., $\{0, 1\}$) has an information content of $I_H = \log 2 = 1$ bit per symbol. When we later apply information theory to physical and chemical systems, we will use the natural logarithm of base e, denoted by ln, yielding the 'natural' information unit often referred to as a 'nat'.

2.1.1 Information and Shannon Entropy

The conceptual expansion of information theory was not undertaken until 1948 when Shannon introduced the foundational information theory that continues to prevail today (Shannon 1948). Subsequently, these concepts found applicability in various scientific domains, including physics; for instance, Brillouin (1956) and Jaynes (1957), which will be elaborated in forthcoming chapters. For a more mathematical exposition of information theory that explicates connections to coding theory, several comprehensive books are available for reference, e.g., Khinchin (1957) and Cover and Thomas (1991).

Shannon considered a situation in which symbols from the alphabet are drawn (or appear) with different probabilities. Initially it is assumed that symbols are uncorrelated. One then finds that the reasonable way to define the *information* $I(p)$ that is gained by observing a symbol—or generally an event—that has an a priori probability p to occur is

$$I(p) = \log \frac{1}{p} \ . \tag{2.3}$$

When making an observation—such as reading the next character in a text or checking the time—information is obtained. The amount of information that is gained from the observation depends on how unexpected the event was. An unexpected or less probable event (small p) corresponds to high information value. In contrast, if we are told something we already know ($p = 1$), the received information is zero.

This definition is a generalisation of Hartley's information quantity as it differentiates between events of varying probability. If the event is an observation of one character from an alphabet consisting of N symbols, where all are assumed to be equally probable, then the probability is $p = 1/N$ and the information is $I(p) = \log N = I_H(N)$.

When one does not have full information about the state of a system, each possible state can be associated with a probability. This implies that the state of the system is described by a probability distribution. This is one of the basic ideas in statistical mechanics, where the precise state (microstate) of a physical system is unknown and hence described as a probability distribution (macrostate) over possible microstates. Such a probability distribution is denoted

$$P = \{p_i\}_{i=1}^{n} \tag{2.4}$$

where p_i denotes the probability for state i, and n is the number of possible states the system can take. The probabilities should be non-negative and normalised

$$p_i \geq 0 \,, \tag{2.5}$$

$$\sum_{i=1}^{n} p_i = 1 \,. \tag{2.6}$$

When one observes a system and learns about its precise state, the amount of information gained depends on which state is observed, as is stated in Eq. (2.3). Consequently, the system can be characterised by the average or expected information one gets when the system is observed. This expectation value, calculated on the basis of the probability distribution P describing the system, is known as the *entropy*, S, of the system:

$$S[P] = \left\langle \log \frac{1}{p_i} \right\rangle = \sum_{i=1}^{n} p_i \log \frac{1}{p_i} . \tag{2.7}$$

(We define $0 \log 0 = 0$, in case p_i equals zero. Throughout the book, we use $< >$ do denote average.) This entropy is usually referred to as the *Shannon entropy*. It is left as an exercise to show that $0 \leq S[P] \leq \log n$.

The entropy is the *expected gain of information* when observing a system characterised by a probability distribution P over its possible states. Alternatively, entropy quantifies the *lack of knowledge* or the *uncertainty* about the system (prior to the observation of the exact state of the system), or the *disorder of the system.*

The maximum value of the entropy S is obtained when all probabilities are equal, $p_i = 1/n$ for $i = 1, ..., n$. This refers to a situation when we have no idea of which state the system is in or what observation we will make. If we know which state is the case, i.e., for a certain i, $p_i = 1$, the entropy is zero, and there is no lack of knowledge. This is illustrated in Fig. 2.1 for a two-state system. The interpretation of entropy as a measure of disorder will be clarified in subsequent chapters as we explore entropies in symbol sequences and apply the information concept to physical and chemical systems.

2.1.1.1 Entropy and Coding—an Extended Example

Consider a stochastic process that generates random sequences of the symbols 'a', 'b', 'c', and 'd', for example 'bcaabada...'. Suppose, to start with, that it is unknown with which probabilities the symbols are generated. Then the best guess is to assign probabilities $1/4$ for each of the events, and that subsequent symbols are independent. The information gained in any of the possible observations of a single symbol is then

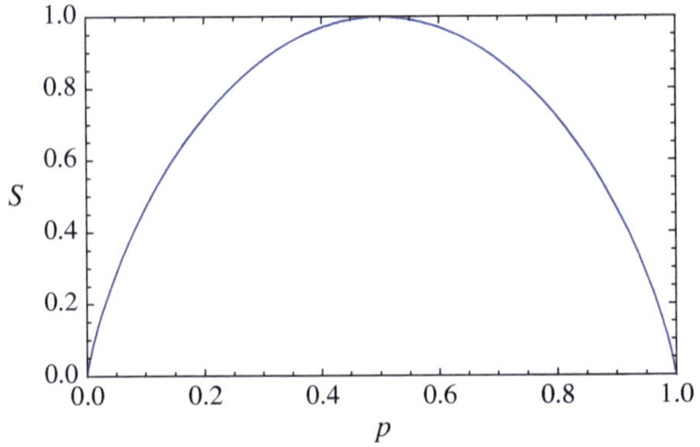

Fig. 2.1 For a system with two possible states ($n = 2$), with the probabilities p and $1 - p$, the entropy $S(p) = S[\{p, 1 - p\}]$, or the lack of knowledge, is at maximum when both states are equally probable ($p = 1/2$). This means that we have no clue on which state we will find the system in when observing it. If we know that the system is in a certain state ($p = 0$ or $p = 1$), then the entropy is zero

$\log 4 = 2$ (bits). This seems reasonable since we can simply code the four symbols with the binary code words '00', '01', '10', and '11', respectively.

Suppose again that symbols are generated independently of each other but that they are generated by the probabilities $p(\text{a}) = 1/2$, $p(\text{b}) = 1/4$, $p(\text{c}) = 1/8$, and , $p(\text{d}) = 1/8$, and that this is known a priori by the observer.

When observing a symbol, the amount of information one gets, according to Eq. (2.1), depends on the symbol. When observing an 'a' we get 1 bit, but a 'd' would give us 3 bits. The expectation value of the information we get from an observation is the weighted average of the information from the four possible events, which is the entropy of Eq. (2.7),

$$S = 1/2 + 1/4 \cdot 2 + 1/8 \cdot 3 + 1/8 \cdot 3 = 7/4. \tag{2.8}$$

We find that the entropy is now reduced, compared to the case with uniform probabilities, since we now have some prior knowledge about the probabilities with which symbols occur. This also means that, by making a better code compared to the trivial one mentioned above, one could compress a message or a sequence of symbols in which the frequencies follow these probabilities. If we for example use the code words '0' for a, '10' for b, '110' for c, and '111' for d, we note that the average code word length decreases from 2 to S above. The knowledge of the frequencies of the four symbols has allowed us to compress the message from 2 bits per symbol to 1.75 bits per symbol. The trick lies in the fact that we have used a code word length (bits in the code word) that is equal to the information gained if the corresponding symbol is observed. So, common symbols that carry little information should be given short code words and vice versa.

2.1.1.2 Entropy as an Additive Quantity

The entropy of a system composed by independent parts equals the sum of the entropies of the parts. The independency means that the probability for a certain microstate of the whole system equals the product of the probabilities of the corresponding microstates of the parts. Assume that the system consists of two subsystems, characterised by the probability distributions

$$Q = \{q_i\}_{i=1}^{n},\; R = \{r_j\}_{j=1}^{m} \tag{2.9}$$

with q_i and r_j representing probabilities for states i and j in the two subsystems, respectively. The whole system is characterised by

$$P = \{q_i r_j\}_{i=1,j=1}^{n,m} \tag{2.10}$$

The entropy can be written

$$\begin{aligned} S[P] &= \sum_{i=1}^{n}\sum_{j=1}^{m} q_i r_j \log \frac{1}{q_i r_j} = \sum_{i=1}^{n} q_i \log \frac{1}{q_i} + \sum_{j=1}^{m} r_j \log \frac{1}{r_j} = \\ &= S[Q] + S[R]\,, \end{aligned} \tag{2.11}$$

where we have used the normalisation, i.e., that $\sum_i q_i = \sum_j r_j = 1$. This shows that, when the subsystems are independent, the entropy of the whole system equals the sum of entropies of its subsystems.

2.1.1.3 Example: Puzzle Solving Using Information

Consider a classic puzzle that can be dealt with using an information-theoretic perspective:

Weighing balls. You have 12 balls, all identical in weight except for one that deviates. Your task is to identify the deviating ball and determine whether it is lighter or heavier using a balance measurement device no more than three times.

To approach this using information theory, start by assessing the initial entropy, denoted by S_0, also reflecting the initial lack of knowledge. To do so, one must consider all possible states of the system with the 12 balls. Any of the 12 balls could deviate, being either heavier or lighter than the others. This results in $n = 12 \cdot 2 = 24$ states, and assuming all states are equally probable, the entropy is at a maximum of $S_0 = \log 24$.

Next, determine how much information is gained from a balance measurement. The quantity of information gained depends on what is placed on the balance. For instance, placing one ball on each side may reveal an unbalanced result, leading to the solution in the next measurement. This would reduce the entropy from log 24 to log 2, providing $\log 24 - \log 2 = \log 12$ bits of information. But it is important to consider the worst-case scenario: a balanced result reduces the number of possible states only from 24 to 22, corresponding to an information gain of $\log 24 - \log 22 \approx 0.126$ bits.

For an optimal measurement arrangement, strive for one where the worst outcome probability is minimised, i.e., the information is maximised. Ideally, the three potential results (tilting left, balanced, tilting right) should be equally probable. In such a scenario, the information gain corresponds to a reduction in the number of possible states by a factor of 3. Observing the balance result has a probability of $p = 1/3$, providing an information gain of log 3. While this ideal measurement setup is not always achievable, log 3 represents the largest information gain we can get in an optimal balance measurement, considering the worst outcome.

This means that with three measurements, we can in the worst case never gain more information than $3 \log 3 = \log 27$. Since this is larger than our initial entropy S_0, it means that it could be possible to solve the problem, provided that we can arrange the measurements sufficiently close to ideal ones.

The first measurement should consist of placing four balls on each side of the balance, leaving the remaining four on the table. The deviating ball is equally likely on the left, on the right, or on the table, so this is an ideal measurement. If one side tilts, say to the left, it indicates that either one of the four balls on the left is heavy, or one of the four balls on the right is light. This situation leaves us with an entropy of $S_1 = \log 8$.

In the second measurement, place three potentially lighter balls aside. On the left side, place two potentially heavier balls and one potentially lighter ball, while the right side holds the remaining two potentially heavier balls and one of known normal weight. While this arrangement is not perfect, it is close enough. The balanced result happens with probability 3/8, tilt left with probability 2/8, and tilt right with probability 3/8. If the balance tilts right, it suggests that either one of the potentially heavy balls on the right is heavy or the potentially light one on the left is light, leaving us with an uncertainty of $S_2 = \log 3$.

Finally, the third measurement can be ideally arranged to eliminate the remaining uncertainty. Simply weigh the two potentially heavy balls against each other to identify the deviant. Figuring out what to do if the first or second measurement results are different from those described is a task left for you to explore.

2.1.2 *Relative Information, Relative Entropy, or Kullback-Leibler Divergence*

Often when we make an observation we do not receive full information of the system. The exact state of the system was possibly not revealed, but based on the observation, we may be able to replace our original, a priori, distribution $P^{(0)}$ with a new one P.

How much information have we gained form this? There are (at least) two approaches for discussing this.

First, we may consider how the "lack of knowledge" of the system has changed. That is quantified by the difference in entropies between the a priori and the new distribution, $S[P^{(0)}] - S[P]$. This is a quantity, though, that can be negative, which would give the picture that the "gained information" is negative. This could happen if we a priori have a skew distribution but learn that a uniform distribution is correct. The entropy difference does indeed capture the difference in knowledge about the system before and after observation, but one usually would say that learning the "correct" distribution is associated with a gain in information. The problem here is connected to the average (or the expectation value) that is calculated in the entropy, and we are comparing quantities using different probabilities for the averaging procedure.

Therefore, in the second approach, we use the new distribution P for the average of both the a priori and the new lack of knowledge. In order to quantify how much information we have gained from the observation, we calculate the decrease in our lack of knowledge of the system. Before the observation, we thought that $p_i^{(0)}$ described the probabilities, but when calculating the expectation value of that lack of knowledge before observation $S^{(0)}$, we should use the new probabilities for the weights,

$$S^{(0)} = \sum_{i=1}^{n} p_i \log \frac{1}{p_i^{(0)}} . \tag{2.12}$$

We assume that whenever a state i is assumed to be impossible in the a priori situation, $p_i^{(0)} = 0$, then this state is also impossible in the new probability distribution, $p_i = 0$.

The lack of knowledge after the observation is the ordinary entropy S, as in Eq. (2.7). The information gained in the observation, the relative information $K[P^{(0)}; P]$, when the a priori distribution $P^{(0)}$ is replaced by the new distribution P, is then

$$\begin{aligned} K[P^{(0)}; P] &= S^{(0)} - S = \\ &= \sum_{i=1}^{n} p_i \log \frac{p_i}{p_i^{(0)}} . \end{aligned} \tag{2.13}$$

Here we define $0 \cdot \log(0/0) = 0$. This quantity is also called the *Kullback-Leibler divergence* (Kullback and Leibler 1951) or the *relative entropy* between the distributions $P^{(0)}$ and P. This quantity fulfils the inequality

$$K[P^{(0)}; P] \geq 0 \tag{2.14}$$

with equality only when the two distributions are identical. This means that the information gained by an observation leading to a change in probabilistic description of the system is always positive. The relative information is 0 only when the a priori description is identical to the new one.

Proof Consider the function $g(x) = x - 1 - \ln x \geq 0$. By adding and subtracting 1, the expression for the relative information may be rewritten as an average over $g(p_i^{(0)}/p_i)$, after transforming from log of base 2 to base e,

$$K[P^{(0)}; P] = \frac{1}{\ln 2} \sum_{i=1}^{n} p_i \ln \frac{p_i}{p_i^{(0)}} = \frac{1}{\ln 2} \sum_{i=1}^{n} p_i \left(\frac{p_i^{(0)}}{p_i} - 1 - \ln \frac{p_i^{(0)}}{p_i} \right) =$$

$$= \frac{1}{\ln 2} \sum_{i=1}^{n} p_i \, g\left(\frac{p_i^{(0)}}{p_i} \right) \geq 0 \,, \tag{2.15}$$

which proves the inequality. It is also clear that equality requires that $p_i^{(0)} = p_i$ for all i.

2.1.2.1 Coding Example of Relative Information

The relative information can be given a coding interpretation as follows. Assume that we have an a priori probability distribution $P^{(0)}$ over 4 symbols, {a, b, c, d} as in the coding example above, and that a stochastic process generates an uncorrelated sequence of those symbols. With the probability distribution as before, $P^{(0)} = \{1/2, 1/4, 1/8, 1/8\}$, we have that the sequence can be ideally compressed from the uncoded 2 bits per symbol in average to $S[P^{(0)}] = 7/4$ bits per symbol. If the distribution of the symbols changes, e.g., to $P^{(0)} = \{1/2, 1/8, 1/4, 1/8\}$, but if we would still use the a priori codes, the average code word length would be $\sum_{\{a,b,c,d\}} p_i \log(1/p_i^{(0)}) = 15/8$ bits. Of course, an optimal code would give the same as before, but since we are using the wrong code with respect to the distribution, we get a less compressed code (15/8 bits vs. 14/8 bits). This difference is also what the relative information quantifies, $K[P^{(0)}; P] = 1/4 \log 2 + 1/8 \log(1/2) = 1/8$ bits. The relative information is thus a measure on the information compression we can get by switching from the wrong to the correct description of the system. In this example, we note that $S[P^{(0)}] = S[P]$, since the new (correct) one is just a permutation of the previous one.

2.1.2.2 The Concavity of Entropy

We shall now use inequality (2.14) to prove that the entropy is a concave function. For the entropy function this means that, if P and Q are two probability distributions (both over n possible states), then the entropy of any weighted average of these is larger than, or equal to, the corresponding weighted average of their respective entropies,

$$S[a \cdot P + (1 - a) \cdot Q] \geq a \cdot S[P] + (1 - a) \cdot S[Q] \,, \tag{2.16}$$

where a and $(1-a)$ are the weight factors $(0 \leq a \leq 1)$. The probabilities in the distributions are denoted p_i and q_i, respectively. The proof is

$$\begin{aligned}
&S[a \cdot P + (1-a) \cdot Q] - (a \cdot S[P] + (1-a) \cdot S[Q]) = \\
&= \sum_{i=1}^{n} (ap_i + (1-a)q_i) \log \frac{1}{ap_i + (1-a)q_i} - \\
&\sum_{i=1}^{n} \left(ap_i \log \frac{1}{p_i} + (1-a)q_i \log \frac{1}{q_i} \right) = \\
&= a \sum_{i=1}^{n} p_i \log \frac{p_i}{ap_i + (1-a)q_i} + (1-a) \sum_{i=1}^{n} q_i \log \frac{q_i}{ap_i + (1-a)q_i} = \\
&= aK[a \cdot P + (1-a) \cdot Q; P] + (1-a)K[a \cdot P + (1-a) \cdot Q; Q] \geq 0\,,
\end{aligned} \tag{2.17}$$

where we have used Eq. (2.14). The non-negative property of the relative information will be used in several proofs for inequalities involving different entropy quantities in later chapters.

2.2 Maximum Entropy Formalism

Even if we do not know exactly the state (microstate) of a certain system, we may have some information on its state. In a physical system, we could, for example, have knowledge about the average energy or the number of particles. Statistical physics is based on the idea that, with such limited information on the state of the system, we make an estimate of the probabilities for the possible microstates. Usually, there is an infinite number of possible probability distributions that are consistent with the known properties of the system in study. The question is then: How should we choose the probability distribution describing our system?

Here it is reasonable to use the concept of entropy, since it can be interpreted as our lack of information on the state of the system. When assigning a probability distribution for the system, we should not use a probability distribution that represents more knowledge than what we already have. For example, when assigning probabilities for the outcome of a roll of a six-sided dice, we say that all probabilities are equal and $1/6$. This corresponds to the situation of maximum entropy, or maximum lack of knowledge. In general we may have some knowledge about the system that implies that this maximum entropy level is not correct. There may be constraints that our knowledge corresponds to which lead to limitations on how the different probabilities can be varied.

Therefore, we choose, among the probability distributions that are consistent with the known system properties, the distribution that maximises the entropy. The probability distribution is in this way derived from a maximisation problem with constraints. This method assures that we do not include any more knowledge, in the description of the system, than we already have. This is the basic idea behind the *maximum entropy principle* (Levine and Tribus 1979), which also can be called the *principle of minimal bias*.

Let us assume that we shall derive a probability distribution $P = \{p_i\}$ describing a system for which we know that the following r averages (or expectation values) hold,

$$\langle f_k \rangle = \sum_{i=1}^{n} p_i f_k(i) = F_k \ , \tag{2.18}$$

for $k = 1, ..., r$. This means that we have r functions $f_k(i), k = 1, ..., r$, of the microstates i, and we know the expectation values, F_k, of these functions. Such a function f could, for example, give the energy of microstate i, or the number of particles found in microstate i, etc. It is now more convenient to work with information based on the natural logarithm (ln). The maximum entropy principle states that we should choose the probability distribution that maximises the entropy under these conditions.

Choose $P = \{p_i\}_{i=1}^{n}$ so that

the entropy $S[P] = \sum_{i=1}^{n} p_i \ln \frac{1}{p_i}$ is maximised, (2.19)

subject to constraints $\langle f_k \rangle = F_k \ , (k = 1, ..., r)$

and the normalisation condition $\sum_{i=1}^{n} p_i = 1$.

We solve this general problem by using the Lagrangian formalism. Therefore, we define the Lagrangian function

$$L(p_1,..., p_n, \lambda_1, ..., \lambda_r, \mu) =$$
$$= S[P] + \sum_{k=1}^{r} \lambda_k \left(F_k - \sum_{i=1}^{n} p_i f_k(i) \right) + (\mu - 1) \left(1 - \sum_{i=1}^{n} p_i \right) , \tag{2.20}$$

where we have introduced Lagrange multipliers (or variables) λ_k, for the r constraints, and $(\mu - 1)$ for the normalisation constraint. (Here we have chosen $\mu - 1$, instead of just μ, because it makes the expressions in the derivation a little simpler.) The optimisation problem is now solved by finding the P, λ_k, and μ that give zeroes in the corresponding partial derivatives of L. (Note that, the idea with the Lagrangian formalism is that the constraints are equivalent to the conditions that the partial derivatives of L with respect to the Lagrange variables are 0.)

Derivation of L with respect to p_j, for any state j, results in

$$\frac{\partial L}{\partial p_j} = -\ln p_j - 1 - \sum_{k=1}^{r} \lambda_k f_k(j) - \mu + 1 , \tag{2.21}$$

which together with the optimisation requirement $\partial L / \partial p_j = 0$ gives

$$p_j = \exp\left(-\mu - \sum_{k=1}^{r} \lambda_k f_k(j)\right) = \exp(-\mu - \lambda \cdot \vec{f}(j)) , \tag{2.22}$$

where we have introduced a vector notation for the Lagrange variables, $\lambda = (\lambda_1, ..., \lambda_r)$, and for the functions f_k, $\vec{f}(j) = (f_1(j), ..., f_r(j))$. A probability distribution of this form is called a Gibbs distribution. The values of the Lagrangian variables λ and μ can be derived from the constraints in Eq. (2.19). Note that the distribution automatically fulfils $p_j \geq 0$. The normalisation condition $\sum_j p_j = 1$ determines μ as a function of λ,

$$\mu(\lambda) = \log Z(\lambda) \tag{2.23}$$

where $Z(\lambda)$ denotes the state sum,

$$Z(\lambda) = \sum_{j=1}^{n} \exp(-\lambda \cdot \vec{f}(j)) . \tag{2.24}$$

Finally, the Lagrange variables λ are determined by the other constraints, Eq. (2.18), which, by use of Eq. (2.23), can be expressed as

$$\frac{\partial \mu}{\partial \lambda_k} = \frac{1}{Z(\lambda)} \sum_{j=1}^{n} (-f_k(j)) \exp(\lambda \cdot \vec{f}(j)) = -\sum_{j=1}^{n} f_k(j) p_j = -F_k , \tag{2.25}$$

for $k = 1, ..., r$. By introducing a vector notation also for constraint values, $\vec{F} = (F_1, ..., F_r)$, this can be written

$$\frac{\partial \mu}{\partial \lambda} = -\vec{F} . \tag{2.26}$$

The maximum entropy value can be expressed in terms of the Lagrange variables as

$$S[P] = \mu + \lambda \cdot \vec{F} . \tag{2.27}$$

(The derivation is left as an exercise.) In the next Section, we illustrate the formalism with an example of a simple Gibbs distribution in physics.

2.2.1 Example: The Bose-Einstein Distribution

Suppose that we have a system composed of a number of particles, where the unknown microstate of the system is described by the exact number of particles n, with n being an integer ($n \geq 0$). Let us assume that we have the knowledge that the expected number of particles in this type of system is N, for example from measurements on a large number of such systems. If this is all we know, the only constraint apart from normalisation is that the expected number of particles equals N. If p_n is the probability that the microstate is composed by n particles, this constraint can be written

$$\sum_{n=0}^{\infty} p_n n = N \ . \tag{2.28}$$

Let λ_1 be the Lagrange variable for this constraint and μ the variable for the normalisation constraint. The state sum, Eq. (2.24), can then be written

$$Z(\lambda_1) = \sum_{n=0}^{\infty} e^{-\lambda_1 n} = \frac{1}{1 - e^{-\lambda_1}} \ . \tag{2.29}$$

By using Eq. (2.25), we find that

$$\frac{\partial \mu}{\partial \lambda_1} = \frac{\partial}{\partial \lambda_1} \ln \frac{1}{1 - e^{-\lambda_1}} = -\frac{e^{-\lambda_1}}{1 - e^{-\lambda_1}} = -N \ , \tag{2.30}$$

which can be used to determine λ_1,

$$\lambda_1 = \ln \frac{N+1}{N} \ , \tag{2.31}$$

and then μ can be derived,

$$\mu = \log Z(\lambda_1) = \ln(N+1) \ . \tag{2.32}$$

The Gibbs distribution is then recognised as the Bose-Einstein distribution,

$$p_n = \frac{N^n}{(N+1)^{n+1}} \ . \tag{2.33}$$

This means that if we know that the expectation value of the number of particles in the system is N, then the least biased probability distribution we can assign for the number of particles in the system is given by p_n in Eq. (2.33).

2.3 Generalisation of Entropies to a Continuous State-Space

The information-theoretic concepts presented above can be generalised to a continuous state-space. Then we assume that we have a probability density $p(\vec{x})$ over this state space, and we suppose that $\vec{x} = (x_1, ..., x_D)$ is a vector in a D-dimensional Euclidean space E. This could, for example, mean that we do not know the exact position of a particle, but that we describe the position of the particle as a probability density over this space. (To be complete and formally correct, we should also introduce a density of states function, but here we assume that to be identical to 1, and therefore we omit it in the expressions.) The probability to find the particle in a certain volume V is then written

$$p(V) = \int_V \mathrm{d}\vec{x}\, p(\vec{x}) \,. \tag{2.34}$$

The probability density is non-negative, $p(\vec{x}) \geq 0$, but has no upper bound. Normalisation requires that integration over the whole space results in unity,

$$\int_E \mathrm{d}\vec{x}\, p(\vec{x}) = 1 \,. \tag{2.35}$$

In analogy with Eq. (2.7), we can now define the entropy S as

$$S[p] = \int_E \mathrm{d}\vec{x}\, p(\vec{x}) \log \frac{1}{p(\vec{x})} \,. \tag{2.36}$$

Like in the discrete case, we can also define a relative information (or a Kullback-Leibler divergence), assuming an a priori probability density $p^{(0)}(\vec{x})$, as a quantification of the information gained when replacing the a priori distribution with a new one $p(\vec{x})$,

$$K[p^{(0)}; p] = \int_E \mathrm{d}\vec{x}\, p(\vec{x}) \log \frac{p(\vec{x})}{p^{(0)}(\vec{x})} \geq 0 \,. \tag{2.37}$$

It is assumed that $p(\vec{x}) = 0$ whenever $p^{(0)}(\vec{x}) = 0$, and we define $0 \cdot \log(0/0) = 0$. One can show that this relative information is non-negative, as in the discrete case. This does not hold for the entropy in Eq. (2.36), though, see Problem 2.10.

In later chapters, we shall use a continuous state-space in the analysis of, for example, spatial structure in chemical systems, in which we represent the spatial concentration distributions of chemical components by probability densities.

Problems

2.1 Balance measurement
Suppose that you have four balls that all look the same, two of equal heavier weight and two of equal lighter weight. Assuming this knowledge, what is the uncertainty of the system. How many measurements using a balance would theoretically be needed in order to sort out which are the heavier and which are the lighter ones? Is there a procedure that accomplishes this?

2.2 Measurement information
Suppose you are to measure the voltage between two points in a circuit, and that you know the value to be anywhere in the range from 0 to 0.5 V (no value is more likely than any other). At your disposal is a Volt meter, showing two significant digits in that range. What is your relative information when you read the result?

2.3 Relative information of a Gaussian
What is the relative information between two Gaussian distributions, with widths b_1 and b_2, respectively? Show that this quantity is always non-negative.

2.4 Information in production error
For a mechanical component to fit in a certain technical system, it is required that its length x fulfils $L - d \leq x \leq L + d$. In the production of these components the resulting lengths are normal (Gaussian) distributed, with an average of L and a standard deviation of $\sigma = d/2$. If we take a new (not yet tested) component and find that it does not fulfil the requirements, what is the relative information?

2.5 A two-step observation
The expected information gain you get when observing the result of throwing a six-sided dice, does not change if you make the observation in two steps. If you first learn whether the result is odd or even, and after that learn the exact outcome, the expectation value of the sum of these two information gains is the same as the expected information gain (which is the entropy) of the direct observation. Show that this results does not depend on the assumption that all outcomes are equally probable, i.e., that this holds for any probability distribution over the outcomes of the dice.

2.6 Maximum entropy
A system is described by a probability distribution over three states, characterised by the energies 0, 1, and 2, respectively. If the expectation value of the energy is 1, what probability distribution should we choose according to the maximum entropy principle?

2.7 Information and decay
Consider a radioactive atom with decay constant λ, so that the probability for the atom to remain after time t is $e^{-\lambda t}$. At time 0 the atom has not decayed. After a certain time $t = 1/\lambda$ the atom is observed again, and it is found to remain in its original state. What is the relative information from that observation?

Now we wait until $t = 2/\lambda$ and observe the atom again. This time it has decayed. What is our information gain in this observation?

2.8 Maximising expected information
Consider again, as in the previous problem, a radioactive atom with decay constant λ. How should we plan to time our measurements if we would like to maximise our expected information gain per observation?

2.9 Maximum entropy of a composite system
A system is composed by n subsystems that each has two possible states, $k_j \in \{0, 1\}$, $(j = 1, ..., n)$, which gives 2^n states for the whole system. Let $K = \sum_j k_j$, and assume that we know that the average of this function is $\langle K \rangle = na$, where $0 \leq a \leq 1$. What is a microstate in this system? What probability distribution over microstates should we choose to describe the system?

2.10 Entropy of Gaussian
What is the entropy of a Gaussian distribution with a width (standard deviation) b? How can this result be interpreted?

2.11 Monty Hall Problem
Behind 1 of 3 closed doors is a prize. You pick one of the doors. Monty opens one of the other doors behind which he knows there is no prize. You are given the choice of sticking with your choice or switching to the other unopened door. Should you switch? What are your chances of winning if you do?

Analyse this in information-theoretic terms. What is your initial uncertainty? How much information do you get when Monty opens a door? How does your probabilistic description change? What is the relative information between your a priori knowledge and the situation after Monty has opened the door?

2.12 Balance information again
You have n balls, all have the same weight except one that deviates.
(a) Use an information-theoretic approach to show that it is not always possible to find the deviating one for $n = 14$ balls.
(b) Show that this is not always possible for $n = 13$ balls unless one is not required to tell whether the deviating one is heavier or lighter.
(c) Tricky: Construct a procedure that finds one ball out of 39 in four balance measurements

2.13 Mathematical requirements for entropy
The quantity $S[P]$ that we have used for entropy of a distribution $P = \{p_1, p_2, ..., p_n\}$ over microstates $(1, 2, ..., n)$ is the only quantity fulfilling the following four conditions: (i) S is symmetric with respect to the probabilities, (ii) S is a continuous function of the probabilities, (iii) The information obtained when one gets to know the outcome of two equally probable events is 1 bit, and finally (iv): The expected gain of information is the same for (I) an immediate observation of the microstate as for (II) a two step observation in which one distinguishes between, say, state 1 and state 2 only if a first observation rejects the other states.

Express the last condition (iv) in mathematical terms, i.e., express the entropy S as a sum of two entropies from two measurements as described in (II). Show that this expression holds for the Shannon entropy, Eq. (2.7).

2.14 Information loss by aggregation
Consider a picture composed of black and white dots, or pixels, in a two-dimensional square lattice of size $N \cdot N$. At the finest resolution, we see exactly where the pixels are and we have full information. Assume that the resolution is made worse (by a factor of two in length scale) so that instead of seeing single pixels, we can only distinguish $(N/2) \cdot (N/2)$ cells, where each cell contains the aggregated information from $2 \cdot 2 = 4$ underlying pixels. At the aggregated level, the cells can take the (observed) "grey-scale" values 0, 1, 2, 3, or 4, corresponding to the number of black pixels that the cell was formed from.

This type of aggregation leads to a loss of information. The loss may vary (locally) depending on the local structure (local densities). Discuss briefly how this works.

If one assumes that the original picture is completely "random", with equal probabilities for black and white pixels, how much information (entropy) do we loose in average (per original pixel) when we aggregate?

2.15 Maximum entropy of particle velocities
Consider a system of uncorrelated particles moving on a one-dimensional lattice. Each particle can have velocity $-2, -1, 0, +1$, or $+2$. The only knowledge you have is that the average velocity is 0 and that the average square velocity is 1, i.e., $\langle v \rangle = 0$ and $\langle v^2 \rangle = 1$. Use the maximum entropy principle to determine the probability distribution over the different velocities.

Chapter 3
Information Theory for Lattice Systems

Abstract The fundamental concepts of information theory, entropy and relative information, can be used to analyse symbol sequences. In such an analysis, the aim is to decompose the information per symbol into two components: the ordered or redundant part of the information content and the disordered part, the entropy. The former quantifies the information present in correlations between symbols, while the latter quantifies the randomness of the sequence. The formalism is illustrated by examples of symbol sequences generated by Markov processes and Hidden Markov models. To characterise the complexity of a symbol sequence, we can use the concept of "effective measure complexity" or "excess entropy". This quantifies one aspect of complexity in the sequence and is explored and discussed in detail. Finally, the chapter concludes by presenting a generalisation of the formalism to two-dimensional patterns of symbols.

3.1 Symbol Sequences and Information

Information theory was developed to analyse signals or sequences of symbols. For instance, if we consider a symbol sequence consisting of the symbols "0" and "1", the information content in each symbol is typically one bit. However, this information quantity can be further analysed and decomposed into two terms. One term quantifies the degree of disorder or randomness in the system, the entropy. The other term measures the degree of order or predictability in the system, which is often referred to as redundancy or superfluous information.

Redundancy arises when there are correlations between symbols in the sequence that allow for predictions about future symbols. For example, by considering the preceding symbols in a sequence, it may be easier to guess the next symbol correctly. The amount of redundancy depends on the strength of the correlations between the symbols.

On the other hand, entropy quantifies the average uncertainty or lack of knowledge per symbol in the sequence, even after accounting for all correlations between symbols. It reflects the level of disorder or randomness in the sequence. In other words, it measures how much we still don't know about the next symbol, even if we

© Springer-Verlag GmbH Germany 2024
K. Lindgren, *Information Theory for Complex Systems*, Understanding Complex Systems, https://doi.org/10.1007/978-3-662-68214-2_3

have taken into account all the information we have so far. It is important to note that the entropy term is where information (in the form of meaningful information) is transmitted between the author and the reader.

Analysis of written texts has for a long time been a popular application of information-theoretic concepts, not the least because of the possibilities to automatically generate texts using a computer program. Randomly generated texts, based on correlation statistics, were present already in Shannon's original paper (Shannon 1948). As we have discussed above, the redundancy in a symbol sequence corresponds to the ordered part of the information content. The disordered information, the entropy, corresponds to the remaining uncertainty, when one is guessing the next symbol in the text, taking into account all correlations. A language with, say, 32 characters (including "space" and some punctuation marks, but without using capital letters) has a maximum entropy value (when correlations are not included) per symbol that is log 32, or 5 bits, but the real entropy is usually much lower. In English about 4 bits are redundant, and only 1 bit is entropy.

How correlations contribute to the structure of the text can be illustrated by random generation of texts, using different correlation lengths. First one needs a large text (or a number of texts), from which statistics is collected on sequences of symbols. The statistics from about a thousand lines of poetry may be sufficient to make the computer seem like an author. If the source for the statistics is only one text written by one author the generated text will not only show the characteristics of the language used, but also the characteristics of the author, as well as features of the story chosen.

Based on statistics from a source text, we form conditional probabilities, $p(x_n|x_1, ..., x_{n-1})$, expressing the probability for the next symbol x_n given the $n - 1$ preceding ones, $x_1, ..., x_{n-1}$. Using these probabilities, we randomly generate symbols, one by one. The longer correlation one decides to include (larger n), the longer preceding sequence is taken into account when calculating the probability. In the following six examples we go from a generated text using only density information (frequencies of symbols) to texts with correlations over block lengths two to six:

1. Tdory d neAeeeko, hs wieadad ittid eIa c i lodhign un a a svmb i ee' kwrdmn.
2. Le hoin. whan theoaromies out thengachilathedrid be we frergied ate k y wee ' e the sle! se at te thenegeplid whe tly titou hinyougea g l fo nd
3. 'Weed. Thed to dre you and a dennie. A le men eark yous, the sle nown ithe haved saindy. If—it to to it dre to gre. I wall much. 'Give th pal yould the it going, youldn't thave away, justove mouble so goink steace, 'If take we're do mennie.
4. I can light,' George tried in you and fire.' 'Nothen it and I want yourse, George some other ther. There's if his hand rolledad ther hisky, 'I little amonely we're we're with him the rain.
5. 'I...I'm not running.' The ranch, work on the time. Do you because you get somethings spready told you just him by heat to coloured rabbits. That's going grew it's like a whisky, place.
6. Million mice because it two men we'll sit by the future. We'll steal it. 'Aren't got it. 'About the fire slowly hand. 'I want, George,' he asked nervoulsly: 'That's fine. Say it too hard forget other.

Already in the case with correlation of two or three, it is quite clear what the language is. When correlation is increased, more of the words are correctly generated, and at the length of six, one may guess from what story the text is taken.

In the following sections, we present a formalism that can be used to analyse disorder and correlations in symbol sequences. In Chap. 4, we use this formalism to analyse states in the form of symbol sequences in the time evolution of discrete dynamical systems (cellular automata). A similar analysis is applied to microstates of a physical system (the Ising spin system model) in Chap. 5. In Chap. 8 on chaotic systems, we illustrate how the formalism can be applied to symbol sequences generated by dynamical systems in order to characterise the dynamics.

It should be noted that there are other approaches for decomposing information in various types of systems. Concepts like 'shared information', 'unique information,' and 'synergistic information' may be parts of such frameworks, see, e.g., Williams and Beer (2010), Bertschinger et al. (2014).

Like in the previous chapter, our definitions on information quantities in symbol sequences depend on the probabilistic description of the system. The formalism builds on the work by Shannon (1948). An analysis of information in correlations was presented by Watanabe (1960) and it was developed in the context of complex systems by several researchers during the 1980s and 1990s, see, for example Grassberger (1986), Eriksson and Lindgren (1987), Lindgren (1987), Lindgren and Nordahl (1988), Crutchfield and Young (1989), Shalizi and Crutchfield (2001), Crutchfield and Feldman (2003).

3.1.1 *Probabilistic Description of a Symbol Sequence*

Let us assume that the system under study consists of an infinite sequence of symbols, in which each symbol is taken from some finite alphabet Λ. In general, the system can be a set of such sequences, an ensemble of possible realisations. We also assume that the system is translation invariant, so that the probability for finding a certain finite sequence of symbols at a certain position in the system does not depend on the position, but only on what other symbols that we may already have observed. In this type of system, we may form probability distributions over sets of finite sub-sequences of symbols, one for each length n (with $n = 1, 2, 3, \ldots$).

Here we will typically consider symbol sequences that are generated by a stationary stochastic process, which implies the translation invariance of the probabilities. Formally, the system can be described by an infinite sequence of stochastic variables X_k,

$$X = (X_0, X_1, X_2, X_3, \ldots) \ , \tag{3.1}$$

and since it is stationary, probabilities for certain sub-sequences $(x_1, x_2, \ldots, x_n)$ do not depend on position,

$$P(X_1 = x_1, X_2 = x_2, ..., X_n = x_n) =$$
$$= P(X_{1+m} = x_1, X_{2+m} = x_2, ..., X_{n+m} = x_n) , \quad (3.2)$$

for all n, m, and any symbols $x_k \in \Lambda$. Therefore we can characterise our system by a probability distribution P_n over symbol sequences of finite length n,

$$P_n = \{p_n(x_1, ..., x_n)\}_{x_1,...,x_n \in \Lambda^n} , \quad (3.3)$$

or shorter

$$P_n = \{p_n(\sigma_n)\}_{\sigma_n \in \Lambda^n} . \quad (3.4)$$

(We are using x, y, z etc. as variables for single symbols and Greek letters σ, α, β etc. as variables for sequences of symbols.) The probabilities fulfil the general requirements of being non-negative and normalised, Eqs. (2.5) and (2.6). Furthermore, there are conditions that relate probability distributions over lengths n and $n + 1$, based on the fact that the distribution over $(n + 1)$-length sequences includes the distribution over n-length sequences. Summation over first or last variable in an $(n + 1)$-length probability results in the corresponding n-length probability,

$$p_n(x_1, ..., x_n) = \sum_{x_0 \in \Lambda} p_{n+1}(x_0, x_1, ..., x_n) , \quad (3.5)$$

$$p_n(x_1, ..., x_n) = \sum_{x_{n+1} \in \Lambda} p_{n+1}(x_1, ..., x_n, x_{n+1}) . \quad (3.6)$$

In the following, we will drop the subscript n of p_n and in short write $p(x_1...x_n)$ or $p(\sigma_n)$, if it is clear which length that is considered.

We usually assume that the stochastic process that generates the sequences of symbols is *ergodic*. This means that, almost always, we can get the right statistics from a single (infinitely long) sequence of symbols $(s_1, s_2, s_3, ...)$ in the ensemble. The *ergodic theorem*, see, e.g., Gray (2009), states that the calculation of an average $\langle f \rangle$ of a function f based on the probabilities $p(x_1...x_n)$ results in the same as an average based on the statistics within the symbol sequence $(s_1, s_2, s_3, ...)$,

$$\sum_{x_1...x_n \in \Lambda^n} p(x_1, ..., x_n) f(x_1, ..., x_n) =$$
$$= \lim_{T \to \infty} \frac{1}{T} \sum_{k=1}^{T} f(s_k, s_{k+1}, ..., s_{k+n-1}) . \quad (3.7)$$

The implication of the *ergodicity property* of a system is that we may consider one outcome of the stochastic process, a specific infinite symbol sequence, and that we may make an internal statistical analysis of it in order to determine a number of information-theoretic properties. Viewed as a physical system, a specific symbol sequence would correspond to a specific microstate. This would then make it possible to quantify entropy and order of a *single microstate*. In statistical physics, this is not the common perspective since one associates the system with the macrostate, or the probability distribution over the possible microstates. We shall return to this view in the chapter on physics and information theory.

3.1.2 Quantifying Disorder in a Symbol Sequence

Let us assume that we have an infinite sequence of symbols (or an ensemble of such sequences), characterised by probabilities for finite length sub-sequences. Our a priori knowledge of the system only reflects that each symbol belongs to a certain alphabet Λ, that contains $|\Lambda| = \nu$ different symbols. Our initial uncertainty (or lack of knowledge) per symbol is then $S = \log \nu$. By successively adding probability distributions for sequences of increasing length P_n (with $n = 1, 2, ...$), we may take correlations into account to reduce our uncertainty of the next symbol in the sequence. The entropy that still may remain when we include all lengths ($n \to \infty$) is the Shannon entropy of the symbol sequence, or for short the *entropy of the symbol sequence*. In some contexts this is called the measure entropy or the entropy rate of the stochastic process (to be discussed in the chapter on chaotic systems).

There are two approaches we can take in order to quantify the randomness per symbol: (i) we can take the entropy of the n-length sequences and divide with the length (in the limit of $n \to \infty$), or (ii) we can quantify the average uncertainty (entropy) of the next symbol, given that we look at all preceding symbols to form the conditional probability for the next. The first approach is straightforward, and is usually used as a definition of the *entropy per symbol*, s. We then need the entropy of n-length subsequences, the *block entropy*, S_m,

$$S_m = S[P_m] = \sum_{x_1...x_m} p(x_1...x_m) \log \frac{1}{p(x_1...x_m)} , \tag{3.8}$$

which are based on the m-length distributions P_m (for all m).

The entropy per symbol s is then defined by

$$s = \lim_{m \to \infty} \frac{1}{m} S_m . \tag{3.9}$$

The second approach (ii) is based on a procedure in which we form conditional probabilities of the next symbol given preceding ones. This approach will also give us tools to quantify how much information that is found in correlations over certain distances in the sequence. This can be derived from the fact that when we extend the preceding sequence that is used to determine the conditional probability for the next symbol, we improve this probabilistic description. The information gained by this improvement, which is based on correlations that are brought in when extending the preceding sequence, can be quantified by the *relative information.*

The conditional probability $p(x_2|x_1)$ for the next symbol, x_2, given a preceding one, x_1, is defined by the probability distributions of 2-length and 1-length sequences, since

$$p(x_2|x_1) = \frac{p(x_1 \text{ and } x_2)}{p(x_1)} = \frac{p(x_1 x_2)}{p(x_1)}, \tag{3.10}$$

where $x_1 x_2$ denotes the pair of the corresponding symbols. In general, the conditional probability $p(x_m|x_1...x_{m-1})$ for a symbol x_m given that we know the preceding $m-1$ symbols $(x_1...x_{m-1})$, is defined by

$$p(x_m|x_1...x_{m-1}) = \frac{p(x_1...x_{m-1}x_m)}{p(x_1...x_{m-1})}. \tag{3.11}$$

This conditional probability is now used to quantify the average uncertainty h_m for the next symbol given a preceding sequence of $m-1$ symbols,

$$\begin{aligned} h_m &= \Big\langle S[p(X_m|x_1...x_{m-1})]\Big\rangle = \\ &= \Big\langle \sum_{x_m} p(x_m|x_1...x_{m-1}) \log \frac{1}{p(x_m|x_1...x_{m-1})} \Big\rangle = \\ &= \sum_{x_1...x_{m-1}} p(x_1...x_{m-1}) \sum_{x_m} p(x_m|x_1...x_{m-1}) \log \frac{1}{p(x_m|x_1...x_{m-1})} = \\ &= \sum_{x_1...x_{m-1}x_m} p(x_1...x_{m-1}x_m) \log \frac{p(x_1...x_{m-1})}{p(x_1...x_{m-1}x_m)}, \end{aligned} \tag{3.12}$$

where we have used the definition of conditional probability, Eq. (3.11). This expression can be rewritten using the block entropies, Eq. (3.8),

$$\begin{aligned} h_m &= \sum_{x_1...x_m} p(x_1...x_{m-1}x_m)\Big(\log \frac{1}{p(x_1...x_{m-1}x_m)} - \log \frac{1}{p(x_1...x_{m-1})}\Big) = \\ &= S_m - S_{m-1} = \Delta S_m, \end{aligned} \tag{3.13}$$

where we have used Eq. (3.6). We note that since entropies are non-negative, $\Delta S_m \geq 0$, the block entropy is an increasing function of block length m. If all correlations (of arbitrary length) to preceding symbols are taken into account when forming the entropy of the conditional probability, this approach should quantify the *entropy of the sequence*, s, as it was defined by Eq. (3.9),

$$\lim_{m\to\infty} h_m = \lim_{m\to\infty} \Delta S_m = \lim_{m\to\infty} \frac{1}{m} S_m = s. \tag{3.14}$$

The second equality, reflecting that both approaches, (i) and (ii), give the same result, will be graphically illustrated from the properties of the block entropies, S_m, in the next section, while a more formal proof is left as an exercise.

3.1.3 *Quantifying Order and Correlations in a Symbol Sequence*

To quantify the information in correlations of length m, suppose that we have an a priori conditional distribution $P^{(0)}(X_m|x_2...x_{m-1})$ for a symbol x_m, given that a specific preceding sequence $(x_2...x_{m-1})$ is known. Then we are interested in the information we get when we observe the symbol x_1, one step further away, and use that to change our probability description of the symbol x_m to $P(X_m|x_1x_2...x_{m-1})$, see Fig. 3.1.

The conditional probabilities in $P^{(0)}$ cannot include any correlations of length m (stretching over a sequence of length m), but that is possible in the new distribution P. The relative information between $P^{(0)}$ and P is then a measure of the correlation information of length m when a specific preceding sequence $(x_1x_2...x_{m-1})$ is observed,

$$K[P^{(0)}; P] = \sum_{x_m} p(x_m|x_1x_2...x_{m-1}) \log \frac{p(x_m|x_1x_2...x_{m-1})}{p(x_m|x_2...x_{m-1})} . \tag{3.15}$$

If we now take the average over all possible preceding sequences $(x_1x_2...x_{m-1})$, we get an expression for the average information content in correlations of length m, the *correlation information* k_m. This quantity can be rewritten in the form of a relative information quantity,

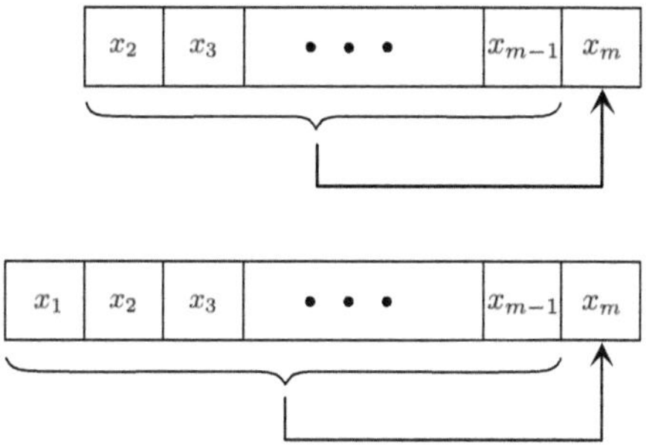

Fig. 3.1 Extending the preceding block of symbols from $(x_2...x_{m-1})$ to $(x_1x_2...x_{m-1})$ allows for longer correlations to be used in the guess of next symbol x_m

$$k_m = \sum_{x_1...x_{m-1}} p(x_1...x_{m-1}) K[P^{(0)}; P] = \tag{3.16}$$
$$= \sum_{x_1...x_{m-1}} p(x_1...x_{m-1}) \sum_{x_m} p(x_m|x_1 x_2...x_{m-1}) \log \frac{p(x_m|x_1 x_2...x_{m-1})}{p(x_m|x_2...x_{m-1})} ,$$

and we note that this quantity is non-negative, $k_m \geq 0$, since it is based on relative information. This can be rewritten as follows, using the definition of conditional probability, Eq. (3.11), and block entropy, Eq. (3.8),

$$k_m = \sum_{x_1...x_{m-1}} p(x_1...x_{m-1}) \sum_{x_m} \frac{p(x_1...x_m)}{p(x_1...x_{m-1})} \log \frac{p(x_1...x_m)p(x_2...x_{m-1})}{p(x_1...x_{m-1})p(x_2...x_m)} =$$
$$= -S_m + 2S_{m-1} - S_{m-2} = -\Delta S_m + \Delta S_{m-1} =$$
$$= -\Delta^2 S_m \geq 0 \, , \tag{3.17}$$

for $m = 2, 3$, etc. Here we define $S_0 = 0$, for Eq. (3.17) to be consistent with Eqs. (3.15)–(3.16).

The non-negativity of the relative information thus implies that $\Delta^2 S_m \leq 0$. This means that the block entropy, S_m, has a positive but decreasing increase with block length m, as is illustrated in Fig. 3.2. From this figure, we can also understand that S_m/m approaches ΔS_m in the limit $m \to \infty$, as stated in Eq. (3.13).

Let us also introduce an information quantity that measures the difference in symbol frequency from a uniform distribution. This is a contribution to the redundant information that is not captured in any correlations. As an a priori distribution we

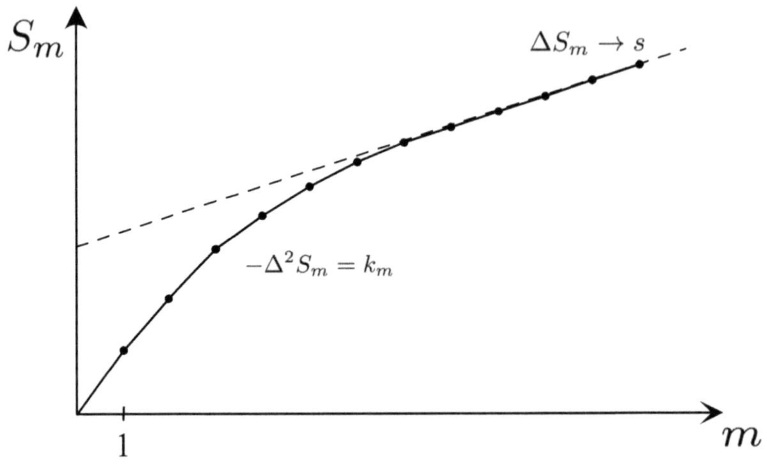

Fig. 3.2 The block entropy S_m is an increasing function of length m. As m increases, the block entropy approaches an asymptotic line with a slope given by the entropy per symbol, s. The decrease in ΔS_m, i.e., $-\Delta^2 S_m = k_m \geq 0$ quantifies the information in correlations k_m over block length m

use the completely "uninformed" uniform distribution $P_1^{(0)}$ that assigns equal probabilities $p_1^{(0)} = 1/\nu$, to all symbols x_1 in Λ. The *density information* k_1 can then be written as the relative information between the a priori uniform distribution and the observed single distribution P_1,

$$k_1 = K[P_1^{(0)}; P_1] = \sum_{x_1} p(x_1) \log \frac{p(x_1)}{1/\nu} = \log \nu - S_1 \ . \tag{3.18}$$

We have thus defined a number of information quantities that capture the redundant (non-random or "ordered") information in the system—the density information k_1 and the series of correlation information contributions k_n ($n = 2, 3, \ldots$). Let us combine all these into an information quantity, the *total correlation information* k_{corr},

$$\begin{aligned} k_{\text{corr}} = \sum_{m=1}^{\infty} k_m = (\log \nu - S_1) + (-S_0 + 2S_1 - S_2) \ + \\ +(-S_1 + 2S_2 - S_3) \ + \\ +(-S_2 + 2S_3 - S_4) \ + \\ \ldots \\ +(-S_{m-2} + 2S_{m-1} - S_m) \ + \\ \ldots = \\ = \log \nu - \lim_{m \to \infty} (S_{m+1} - S_m) = \log \nu - s \ . \end{aligned} \tag{3.19}$$

This is the redundant information in the system, expressed as an average per symbol. The entropy per symbol of the system, s, is a measure of the uncertainty that remains when all correlations, including deviation in symbol frequencies from uniformity, have been taken into account.

From this we can conclude, that the total entropy per symbol of $\log \nu$, which is the maximum entropy per symbol, can be decomposed in two terms, the redundancy k_{corr} and the entropy s,

$$S_{\max} = \log \nu = (\log \nu - s) + s = k_{\text{corr}} + s \ , \tag{3.20}$$

where the correlation information has one term expressing density information, k_1, and a series of terms, k_m, quantifying information in correlations of different lengths,

$$k_{\text{corr}} = k_1 + \sum_{m=2}^{\infty} k_m \ . \tag{3.21}$$

The interpretation of these quantities in terms of coding theory is the following. If a text contains redundancy this can be removed in order to compress the text by using a coding procedure, replacing symbols in an appropriate way. In the case of an optimal code, the compressed sequence may have a length of $Ls/\log \nu$ if the original length is L.

3.2 Markov Processes and Hidden Markov Models

3.2.1 Markov Processes and Entropy

Some of the symbol sequences analysed in later chapters are the direct result of simple Markov processes, i.e., processes in which the probability for the next state (or symbol) is fully determined by the preceding state (or symbol). This means that the conditional probability distribution $P(Z_n|z_1...z_{n-1})$ over the next symbol z_n, given a preceding sequence of symbols, converges already for $n = 2$. The entropy s of the process is then given already by ΔS_2, see Eqs. (3.12)–(3.13),

$$s = \lim_{n\to\infty} \sum_{z_1...z_{n-1}} p(z_1...z_{n-1}) \sum_{z_n} p(z_n|z_1...z_{n-1}) \log \frac{1}{p(z_n|z_1...z_{n-1})} =$$

$$= \sum_{z_{n-1}} p(z_{n-1}) \sum_{z_n} p(z_n|z_{n-1}) \log \frac{1}{p(z_n|z_{n-1})} = \Delta S_2 \; . \tag{3.22}$$

A Markov process can be described as a finite automaton with N internal states z_i (with $i = 1, ...N$, and z_i belonging to the alphabet Λ) corresponding to the symbols generated. The process changes internal state according to transition probabilities, P_{ij}, denoting the probability to move to state j from state i. An example of such an automaton is given in the figure below. The internal states (the possible symbols) are a, b, and c. Let us assume in this example that, when there is a choice for the transition from a state, all possible transitions from that state are equally probable. Here this means that $P_{\mathrm{aa}} = P_{\mathrm{ac}} = P_{\mathrm{ba}} = P_{\mathrm{bc}} = 1/2$, $P_{\mathrm{cb}} = 1$, and $P_{\mathrm{ab}} = P_{\mathrm{ca}} = P_{\mathrm{cc}} = P_{\mathrm{bb}} = 0$.

To calculate the entropy of the process (or the symbol sequence), we need $p(z)$, which is the probability distribution over the internal states (or, equivalently, the stationary probability distribution over the internal states). Since the (stationary) probability $p(z)$ to be in a certain state z must equal the sum over the probabilities of possible preceding states w weighted with the transition probabilities to state z, we can determine $p(z)$ by

$$p(z) = \sum_{w} p(w) P_{wz} \; , \tag{3.23}$$

for all $z \in \Lambda$, together with the normalisation constraint $\sum_z p(z) = 1$. Note that the transition probabilities P_{zw} equals the conditional probabilities $p(w|z)$, which means that the entropy s can be written

$$s = \sum_z p(z) \sum_w P_{zw} \log \frac{1}{P_{zw}} . \tag{3.24}$$

In the example above, the stationary distribution over the states is found by solving Eq. (3.23), replacing, e.g., the last one with the normalisation constraint,

$$\begin{aligned} p(\mathtt{a}) &= \frac{1}{2}\, p(\mathtt{a}) + \frac{1}{2}\, p(\mathtt{b}) , \\ p(\mathtt{b}) &= p(\mathtt{c}) , \\ p(\mathtt{c}) &= 1 - p(\mathtt{a}) - p(\mathtt{b}) , \end{aligned} \tag{3.25}$$

and one finds that $p(\mathtt{a}) = p(\mathtt{b}) = p(\mathtt{c}) = 1/3$. The term $-P_{zw} \log P_{zw}$ only contributes to the entropy in Eq. (3.24) when z is a or b, so the resulting entropy then turns out to be $s = 2/3$ (bits). One way to view this is that the process needs one bit of "random information" in order to make the random choice of transition when leaving node a and node b. Since these nodes are visited a fraction of 2/3 of the time, and since this choice directly determines the symbols in the sequence, the average randomness, or the entropy, must be 2/3 bits.

3.2.1.1 An Example of an Optimal Code Exploiting Correlations

Consider a stochastic process producing sequences of symbols from the alphabet $\Lambda = \{\mathtt{a}, \mathtt{b}, \mathtt{c}, \mathtt{d}\}$, like in Example 2.1.1.1. Now we assume that there are correlations between neighbouring symbols. If the preceding symbol is observed, we have some additional knowledge about the probability for the following one.

The correlations between the symbols are determined by the structure of the Markov process generating the symbol sequence. The finite state automaton in Fig. 3.4 defines the Markov process. Solving for the stationary distribution, Eq. (3.23), results in the single symbol distribution $P_1 = \{1/2, 1/4, 1/8, 1/8\}$, as we also had in the Example 2.1.1.1. The entropy s of the process is then determined by Eq. (3.22). This implies that there is a single bit of contribution to the entropy each time the process leaves any of the states a or b (since there is always 1/2 chance for choosing a specific transition), but no contribution when leaving states c or d. The entropy s is then 1/2 + 1/4 bits = 3/4 bits. This is significantly lower than the 7/4 bits that we found in Example 2.1.1.1, and what we would also get here if we would disregard correlations, i.e., the entropy of the single symbol distribution P_1.

The implication of this is that we should be able to find an even more efficient coding, compressing the length of a binary coded message down to the entropy of 3/4 bits in average per original symbol. One easy way to achieve this is to start with

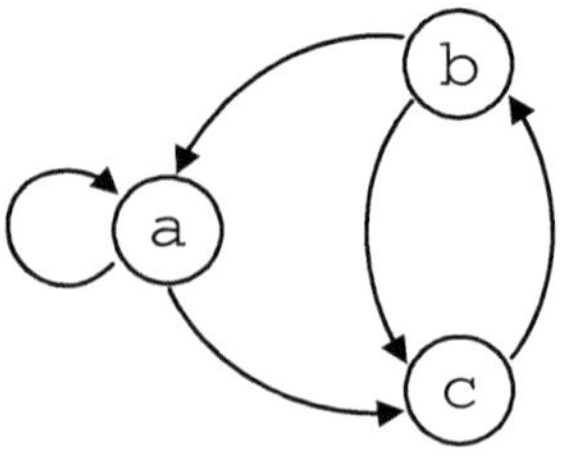

Fig. 3.3 Example of finite state automaton representing a Markov process

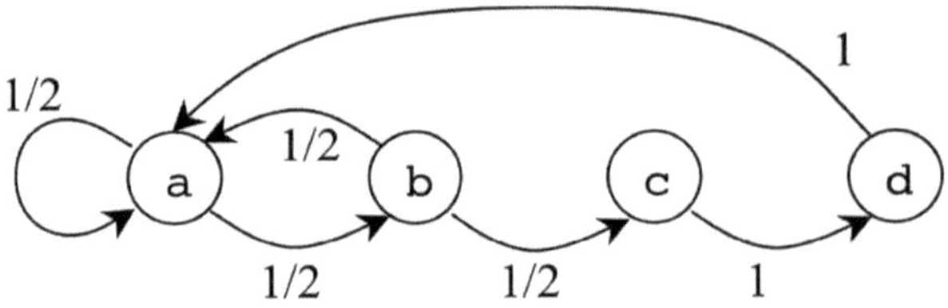

Fig. 3.4 Markov process generating sequences of symbols from the alphabet $\{a, b, c, d\}$. When two arcs leave a node, the probability for choosing an arc is $1/2$, as indicated on the arcs

the code word for the initial symbol using the code of Example 2.1.1.1. After that the code words are 0 or 1 only, for states *a* and *b*, with 0 meaning a transition to a and 1 for the other transition. For states *c* and *d* no code word is needed as the next state is unique.[1] Then, in average for a long message (in the infinity limit), we only need one bit after state *a* and another bit after state *b*, which occur in 3/4 of all positions in the original sequence, resulting in an average code word length of 3/4.

3.2.2 *Hidden Markov Models and Entropy*

For some stochastic processes, the observation of a generated symbol may not reveal the internal state of the system, as it does for a Markov process. In a *Hidden Markov model*, one does not observe the states z of the process (or of the finite state automaton), but one observes some function of the state $f(z)$. If the function f is not invertible this means that information may be lost between the underlying process and the sequence of symbols that is observed. In the example of Fig. 3.3, if the function f is given by $f(\texttt{a}) = 0$ and $f(\texttt{b}) = f(\texttt{c}) = 1$, then we have a process that

[1] In order to make sure one knows where the end of the message is (since there is no code word following states *c* and *d*), one may encode the last symbol of the message using the reversed code words that were used for the initial symbol (so that one can read the last code word from right to left at the end).

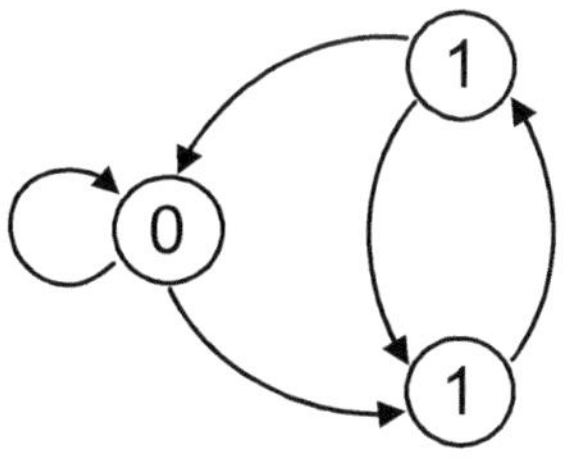

Fig. 3.5 Example of finite state automaton representing a Hidden Markov model based on the Markov process of Fig. 3.3

generates a sequence of '0' and '1' symbols, which is not a Markov process but a Hidden Markov model. The process can now be illustrated by the automaton in Fig. 3.5.

The process thus generates sequences of 0's and 1's in which the restriction is that blocks of 1's (separated by 0's) always are of even length. Since we only observe the generated symbols 0 or 1, we do not always immediately know which state the automaton is in. But if we are allowed to observe an arbitrarily long sequence of generated symbols, we will *almost always* find a 0 in the preceding sequence.[2] And then we can trace the path through the automaton to determine in which node we are, which then gives us the true probabilities for the next symbol. This procedure works for this specific example, but that does not apply in general.

Using Eqs. (3.12)–(3.13), we can write the entropy

$$s = \lim_{n\to\infty} \sum_{x_1...x_{n-1}} p(x_1...x_{n-1}) \sum_{x_n} p(x_n|x_1...x_{n-1}) \log \frac{1}{p(x_n|x_1...x_{n-1})} =$$

$$= \lim_{n\to\infty} \sum_{z} \sum_{(x_1...x_{n-1})\to z} p(x_1...x_{n-1}) \sum_{x_n} p(x_n|z) \log \frac{1}{p(x_n|z)} =$$

$$= \sum_{z} p(z) \sum_{x} p(x|z) \log \frac{1}{p(x|z)} , \tag{3.26}$$

where we note that the sum of the probabilities of all preceding sequences ending in node z, denoted $(x_1...x_{n-1}) \to z$, is the stationary probability $p(z)$ for being in node z, and $p(x|z)$ denotes the probability of generating symbol x when in node z. In this case, it thus turns out that the entropy generated by the underlying Markov process, i.e., the entropy in the choices on the level of the automaton, can actually be detected in the Hidden Markov model process, in which only the symbol sequences of 0's and 1's are observed. We do need to take into account longer correlations, though, in order to find the entropy. This implies that this Hidden Markov model has a different correlation information distribution than the corresponding Markov process. This example is further discussed in the next section, illustrating also correlation information quantities.

[2] Only if the preceding sequence is an infinite sequence of 1's, we cannot say where we are.

3.2.3 Some Examples

In the following four examples, the processes that generate the symbol sequences are expressed with finite state automata in which the symbols are on the transition arcs instead of in the nodes (internal states), as we have exemplified above. Make sure that you understand how you can move between these two representations.

3.2.3.1 Example: Crystal

Consider a periodic symbol sequence of 0's and 1's,

...01010101010101010101010101010101...

The density information is zero, $k_1 = 0$, since the probabilities are equal for the two symbols, $p(0) = p(1) = 1/2$, and they do not deviate from a uniform a priori distribution. The probabilities that are required for calculating the correlation information are

$$p(0) = p(1) = \frac{1}{2}\,, \quad p(0|1) = p(1|0) = 1\,, \quad \text{and } p(0|0) = p(1|1) = 0\,. \tag{3.27}$$

The correlation information from length $n = 2$, calculated by Eq. (3.16), then gives

$$k_2 = \sum_{x_1} p(x_1) \sum_{x_2} p(x_2|x_1) \log \frac{p(x_2|x_1)}{p(x_2)} = \log 2 = 1 \text{ (bit)}. \tag{3.28}$$

Since this is the total information (per symbol) of the system, we can conclude that all other redundant information quantities are zero, $k_m = 0$ for $m \neq 2$, and $s = 0$. This is what one should expect. There is no entropy in this system—as soon as we see one symbol ("0" or "1"), we know the next symbol, and so on.

3.2.3.2 Example: Gas

Consider instead a symbol sequence generated by a completely random process, like coin tossing,

...110000110100010110010011011101...

The probability for next symbol to be "0" or "1" is 1/2 independently of how many preceding symbols that we may observe. The entropy of the conditional probability is therefore always maximal, $\log 2 = 1$ (bit),

$$s = 1 \text{ (bit)} \tag{3.29}$$

and there are no contributions from the redundancy, $k_{\text{corr}} = 0$.

3.2.3.3 Example: Finite State Automaton Generating Short Correlations

Let us now consider a symbol sequence generated by a stochastic process described by a finite automaton, see Fig. 3.6. Here we have used a representation where symbols are generated on the transition arcs instead of in the nodes. (It is straightforward to convert from one representation to the other.) This automaton generates symbol sequences where 0's cannot appear in pairs, e.g.,

...1101011111011110101011101110110...

In order to calculate the information-theoretic properties, we need to transform the characteristics of the automaton to a probabilistic description of the symbol sequence. First, we calculate the densities of 0's and 1's. Since the transition (arc) to the right node (R) always generates a "0" and is the only way a "0" can be generated the probability for being in R, p(R) equals the frequency of 0's, $p(0) = p(\text{R})$. Similarly, $p(1) = p(\text{L})$. The probabilities for the nodes are given by the stationary probability distribution over the nodes. Here, this can be expressed by the fact that the probability for being in the left node p(L) must be equal to the probability that we were in this node last step and generated a "1" plus the probability that we were in the right node last step,

$$p(L) = p(L)\frac{1}{2} + p(R) \;\Rightarrow\; p(L) = \frac{2}{3} \text{ and } p(R) = \frac{1}{3}\,, \tag{3.30}$$

where we have used the normalisation $p(\text{L}) + p(\text{R}) = 1$. This means that $p(0) = 1/3$ and $p(1) = 2/3$. Then the density information is

$$k_1 = \sum_{x_1} p(x_1) \log \frac{p(x_1)}{1/2} = \frac{5}{3} - \log 3 \approx 0.0817 \;\; \text{(bits)} \tag{3.31}$$

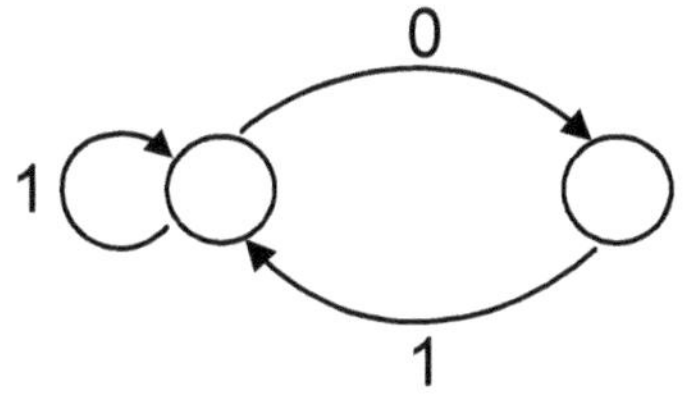

Fig. 3.6 The finite automaton generates an infinite sequence of "0" and "1" by following the arcs between the nodes in the graph. When there are two arcs leaving a node, one is chosen randomly with equal probabilities for both choices. Note that we can easily transform the automaton to a representation with symbols in the nodes by having 1's generated by the left node and 0's by the right one

All other redundant information is contained in correlations over block length $n = 2$, since if we observe one symbol we know which node in the automaton that is the starting point for generating the next symbol, and then we have the full knowledge about the true probabilities for that symbol. Thus, $p(0|1) = p(1|1) = 1/2$, $p(1|0) = 1$, and $p(0|0) = 0$. The correlation information for length two is then

$$k_2 = \sum_{x_1} p(x_1) \sum_{x_2} p(x_2|x_1) \log \frac{p(x_2|x_1)}{p(x_2)} = \log 3 - \frac{4}{3} \approx 0.2516 \text{ (bits)} \quad (3.32)$$

The entropy s is the entropy (uncertainty) about the next symbol in the sequence, based on our knowledge on all preceding symbols. The preceding symbols inform us on which node is used in generating the next symbol, and, actually, this information is in the last symbol alone. Formally, this can be expressed

$$p(1|x_1...x_{n-1}1) = p(0|x_1...x_{n-1}1) = \frac{1}{2}\,, \quad \text{for all possible } x_1...x_{n-1}\,, \text{ and}$$
$$p(1|x_1...x_{n-1}0) = 1\,, \quad \text{for all} \quad \text{possible } x_1...x_{n-1}\,.$$

It is clear that we have a Markov process, and therefore Eq. (3.22) can be used to calculate the entropy,

$$s = \lim_{m\to\infty} \Delta S_m = \Delta S_2 = \sum_{x_1} p(x_1) \sum_{x_2} p(x_2|x_1) \log \frac{1}{p(x_2|x_1)} =$$
$$= p(1) \cdot \log 2 + p(0) \cdot 0 = \frac{2}{3} \text{ (bits)} \approx 0.6667 \text{ (bits)} \quad (3.33)$$

This is, of course, what we should expect, since we have already said that there is no more correlation information than what was calculated in k_1 and k_2. What remains when the redundancy has been subtracted from the total of 1 bit of information per symbol must be the entropy, i.e., $s = 1 - (5/3 - \log 3) - (\log 3 - 4/3) = 2/3$.

3.2.3.4 Example: Finite State Automaton Generating Long Correlations

The finite automaton of Fig. 3.7 is similar to the one of the previous example, with the difference that the arc leading from the right to the left node generates a "0" here. This means that the automaton generates sequences where 1's are separated by an even number of 0's. This is a Hidden Markov model, similar to the one we discussed in the example of Sect. 3.2.2.

Suppose that we shall guess on the next symbol in the sequence, denoted by x, and that we may take into account an large (infinite) number of preceding symbols, for example,

$$...0001000011100100000000110000000\,x \quad (3.34)$$

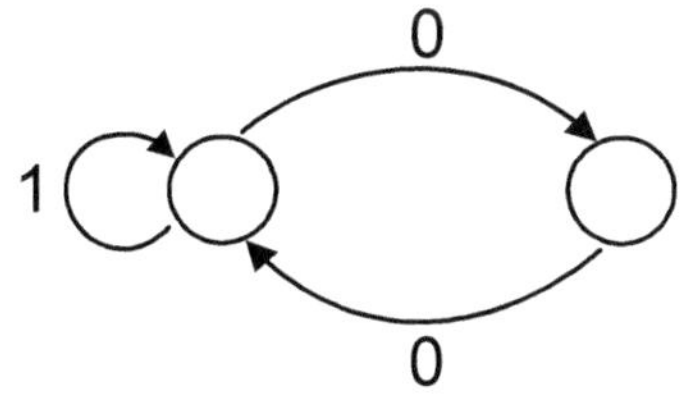

Fig. 3.7 Finite automaton generating sequences of 0's and 1's in which 1's are always separated by an even number of 0's

Then it is sufficient to go back to the nearest preceding "1" and count how many 0's there are in between. If there is an even number (including zero 0's), we are in the left node, and if there is an odd number, we are in the right node. When we know which node we are in, we also have the true probability description of the next symbol. (Only if there are only 0's to the left, no matter how far we look, we will not be able to tell which is the node, but as the length of the preceding sequence tend to infinity the probability for this to happen tends to zero.)

Since the preceding sequence almost always determines (and corresponds to) the actual node, in the limit of infinite length, we can rewrite the entropy s as follows.

$$\begin{aligned} s &= \lim_{n\to\infty} \sum_{x_1 \ldots x_{n-1}} p(x_1 \ldots x_{n-1}) \sum_{x_n} p(x_n | x_1 \ldots x_{n-1}) \log \frac{1}{p(x_n | x_1 \ldots x_{n-1})} = \\ &= p(L) \sum_x p(x|L) \log \frac{1}{p(x|L)} + p(R) \sum_x p(x|R) \log \frac{1}{p(x|R)} = \\ &= \frac{2}{3} \log 2 + \frac{1}{3} \cdot 0 = \frac{2}{3} \text{ (bits)} \approx 0.6667 \text{ (bits)} \end{aligned} \tag{3.35}$$

We have used the probabilities for the nodes L and R from the previous example, since they are the same. Expressed in this way, it is clear that the entropy of the symbol sequence comes from the random choice the automaton has to make in the left node. The right node does not generate any randomness or entropy. Therefore, we get the same entropy as in the previous example. Also the frequency distribution of single symbols are the same (but with shifted order between 0 and 1) resulting in the same density information k_1, but when we get to the correlation information over blocks, we get a difference. In this case, we have correlation information in arbitrarily large blocks, since any number of consecutive 0's may occur, and then there is more information to be gained by observing one more symbol. To calculate the general expressions for the different correlation information terms is now much more complicated, and we leave it as a difficult exercise.

So, even if the last two discussed examples are identical in terms of entropy and redundancy, this second example may be considered more complex, since the correlation information is spread out over larger distances. In the next section, we shall see that such a difference may be used as one way to characterise the complexity of the symbol sequences.

3.3 Measuring Complexity

What characterises a complex symbol sequence or a complex pattern? There are numerous suggestions on how one should quantify complexity. What quantity to use depends on what one is looking for in the system under study. In this section we will focus on quantities related to how correlation information is distributed in the system. This approach was suggested by Peter Grassberger in the 1980s, see Grassberger (1986). Consider two of the examples discussed above, the completely ordered ("crystal") sequence and the completely random ("gas") sequence:

...01010101010101010101010101010...

...11000011010001011001001101110 1...

In the first case, the entropy is minimal (0 bit), and in the second case it is at maximum (1 bit). None of these are typically considered as complex, even though, in some contexts, the random one has been called complex.[3] The precise configuration of the gas needs a lot of information to be specified, but the ensemble or the system that generate that type of sequence is simple: "toss a coin forever...". The crystal has a simple description as it is: "repeat 01...".

If neither the most ordered nor the most random systems are considered complex, we should look for a measure of complexity that can take high values when the entropy is somewhere in between, see Fig. 3.8. One information-theoretic characteristic that has been considered important for a complex system is to what extent correlation information stretches over large distances in the system. If there is a lot of information in long-range correlations, it may be more difficult to analyse and describe the system. Grassberger (1986) suggested a quantity, *the effective measure complexity*, η, which can be defined as a weighted sum of correlation information contributions from different block lengths,

$$\eta = \sum_{m=1}^{\infty} (m-1) k_m \, . \tag{3.36}$$

This quantity has also been introduced and investigated by Jim Crutchfield and coworkers under the term *excess entropy*, see, e.g., Crutchfield and Young (1989), Shalizi and Crutchfield (2001), Crutchfield and Feldman (2003). We call this the *correlation complexity*, emphasising the correlation quantification, and it can be rewritten (if $k_{\text{corr}} > 0$) as

$$\eta = k_{\text{corr}} \sum_{m=1}^{\infty} (m-1) \frac{k_m}{k_{\text{corr}}} = k_{\text{corr}} \overline{(m-1)} = k_{\text{corr}} \, d_{\text{corr}} \, . \tag{3.37}$$

[3] In algorithmic information theory, high 'complexity' has been associated with a system that requires a long description to reproduce it in detail, see Kolmogorov (1965), Chaitin (1966).

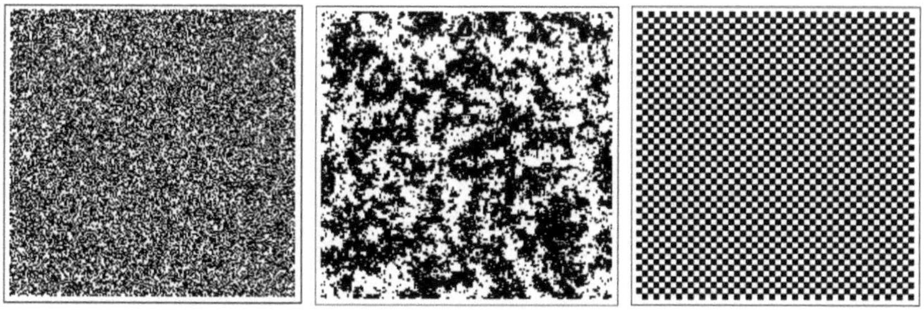

Fig. 3.8 One is usually not considering a fully random pattern (left) as complex. Neither is a fully ordered structure, exemplified by the checkerboard pattern (right), considered as complex. The potential to exhibit complex characteristics usually requires patterns in between these extremes—patterns that may contain some order in correlations, but where correlations may stretch over larger distances, and where there is also some randomness as well. Here this is illustrated by a typical state in a physical spin system close to the critical temperature (middle)

Here we have introduced an average correlation distance d_{corr}. This is based on the average block length at which correlation information is found, but where correlation distance is defined to be one less than the block length (or the "distance" between first and last symbol of the block). This complexity measure is non-negative and there is no upper bound for it. For the "crystal" example $\eta = 1$ and for the totally random sequence $\eta = 0$.

It also turns out that the correlation complexity can be interpreted as the average information contained in a semi-infinite symbol sequence $(...x_{-2}, x_{-1}, x_0)$ has about its continuation $(x_1, x_2, x_3, ...)$. This quantity can be written as a relative information. Suppose first that the preceding sequence, that we may observe, has a finite length m and denote it σ_m. The continuation, has finite length n and we denote it τ_n. At the end we shall look at infinite limits of m and n. The a priori description of the continuation is given by probabilities $p(\tau_n)$, but after we have observed the preceding sequence σ_m, we can replace that with the conditional probabilities $p(\tau_n|\sigma_m)$. The information we gain by this is a relative information quantity, and we take the average over all possible preceding sequences together with the limit of infinite lengths,

$$\eta = \lim_{m\to\infty} \lim_{n\to\infty} \sum_{\sigma_m} p(\sigma_m) \sum_{\tau_n} p(\tau_n|\sigma_m) \log \frac{p(\tau_n|\sigma_m)}{p(\tau_n)} \geq 0\,. \tag{3.38}$$

One can also relate the correlation complexity to the block entropies, see Fig. 3.9,

$$\eta = \lim_{m\to\infty} (S_m - ms)\,, \tag{3.39}$$

where s is the Shannon entropy of the symbol sequence (or more correctly of the stochastic process). The correlation complexity can thus be interpreted as the rate of convergence of S_m/m to s. The proof that these expressions are equal to the correlation complexity of Eq. (3.36) is left as an exercise.

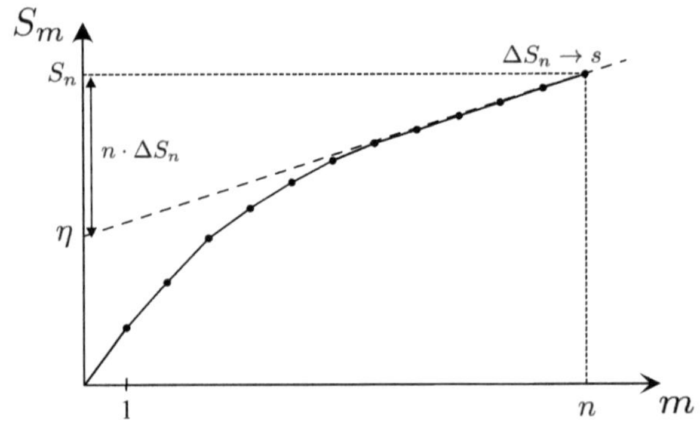

Fig. 3.9 The block entropy S_n approaches an asymptotic line with slope equal to the entropy s, as block length n increases. Also the ratio S_n/n converges to the entropy, but more slowly. Hence, in the limit of infinite block length, $S_n - ns$ is a measure of how fast S_n/n converges to the entropy s. This is the correlation complexity η, and graphically, it is the intersection point on the vertical axis of the asymptotic line that is approached by the block entropy

3.3.1 *Correlation Complexity for Markov Processes and Hidden Markov Models*

For a Markov process the correlation complexity η is easy to calculate since there is no correlation information from blocks longer than 2, and therefore we get that $\eta = k_2$. For a Hidden Markov model, the situation is different. As we saw in the example of Sect. 3.2.3.4, such a system may have correlation information in arbitrarily long blocks of symbols.

In some situations there may be an easy way to calculate the correlation complexity using the form that expresses the information contained in the past about the future of an infinite symbol sequence, Eq. (3.38). Consider again the automaton describing the process of example Sect. 3.2.3.4, see Fig. 3.10.

If almost all past sequences σ_m, in the infinite limit, determine whether we are in the left (L) or in the right (R) node, we can rewrite Eq. (3.38). The probability of the

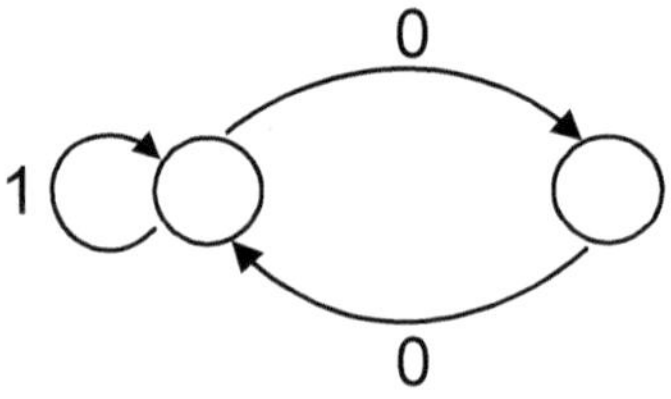

Fig. 3.10 Example of Hidden Markov model, represented by finite automaton generating sequences of 0's and 1's in which 1's are always separated by an even number of 0's. The probability for the arcs leaving the left node are both 1/2

future sequence τ_n is given by the node we are in, so we can group together all σ_m leading to the left node L (and in the same way all leading to the right one, R). Then we have

$$\eta = \lim_{n\to\infty} \sum_{z\in\{L,R\}} p(z) \sum_{\tau_n} p(\tau_n|z) \log \frac{p(\tau_n|z)}{p(\tau_n)} , \tag{3.40}$$

where $p(L)$ and $p(R)$ are the stationary probabilities for the left and right node, respectively. We now use the definition of conditional probability $p(\tau_n|z)$ for the n-length sequence τ_n given that we are in node z, to rewrite the argument in the logarithm,

$$\frac{p(\tau_n|z)}{p(\tau_n)} = \frac{p(\tau_n, z)}{p(\tau_n)p(z)} = \frac{p(z|\tau_n)}{p(z)} . \tag{3.41}$$

Then we can rewrite Eq. (3.40) as follows,

$$\begin{aligned}
\eta &= \lim_{n\to\infty} \sum_{z\in\{L,R\}} p(z) \sum_{\tau_n} p(\tau_n|z) \log \frac{p(z|\tau_n)}{p(z)} = \\
&= \lim_{n\to\infty} \sum_{z\in\{L,R\}} p(z) \sum_{\tau_n} p(\tau_n|z) \left(\log \frac{1}{p(z)} + \log p(z|\tau_n) \right) = \\
&= \sum_{z\in\{L,R\}} p(z) \log \frac{1}{p(z)} - \lim_{n\to\infty} \sum_{\tau_n} p(\tau_n) \sum_{z\in\{L,R\}} p(z|\tau_n) \log \frac{1}{p(z|\tau_n)} .
\end{aligned} \tag{3.42}$$

The last sum over nodes z is the entropy of which node we were in conditioned on observing the future sequence τ_n. But, in this example, the starting node for the future sequence is almost always uniquely given by the sequence τ_n. Only when there are only zeroes in τ_n, we do not know the starting node, but that happens with probability 0 in the infinite n limit. Therefore there is no uncertainty of z given the future τ_n, and the last entropy term is 0. Consequently, in this case,

$$\eta = \sum_{z\in\{L,R\}} p(z) \log \frac{1}{p(z)} . \tag{3.43}$$

We conclude that, for this situation, the correlation complexity equals the entropy of the stationary distribution over the nodes in the finite state automaton describing the Hidden Markov model. Note, though, that this does not hold in general, but only for certain types of Hidden Markov models.

3.4 Extensions to Higher Dimensions

The decomposition of information into entropy and redundancy with contributions from different correlation lengths can be extended to lattice systems of any dimension. In this section we briefly show how the extension to the two-dimensional case is done, but it is easy to generalise the formalism also to higher dimensions. The approach presented here (Lindgren 1989), in which one "reads" symbols sequentially was exploited already by Kramers and Wannier (1941), as a direct generalisation from one dimension. In fact, there is an infinity of ways to derive the entropy of a lattice system not necessarily using a sequential "reading" of the symbols, see the work by Helvik and Lindgren (2014).

Consider an infinite two-dimensional lattice in which each site is occupied by a symbol (0 or 1). (The generalisation to larger alphabets is straightforward.) Assume that the relative frequencies, with which finite configurations of symbols occur in the lattice, are well defined. Let $A_{M \times N}$ be a specific $M \times N$-block occurring with probability $p(A_{M \times N})$. Then, in analogy with the one-dimensional symbol sequence, Eq. (3.9), the entropy, i.e., the average randomness per site, can be defined by

$$s = \lim_{M,N \to \infty} \frac{1}{MN} S_{M \times N} \,, \tag{3.44}$$

where the block entropy $S_{M \times N}$ is defined by

$$S_{M \times N} = \sum_{A_{M \times N}} p(A_{M \times N}) \log \frac{1}{p(A_{M \times N})} \,. \tag{3.45}$$

In order to construct an alternative way to derive the entropy per site, based on a sequential process that reads the states (or symbols) in the pattern one by one, cf. Eq. (3.12), we introduce a block configuration of a shape that makes such an approach possible. Let B_m be a certain configuration of symbols arranged as follows: $m - 1$ rows of symbols, each of length $2m - 1$, are put on top of each other, and on the $m - 1$ first symbols of the top row a sequence of $m - 1$ symbols is placed, see Fig. 3.11. We also introduce the notation $B_m x$ for the configuration that adds the symbol x to B_m after the $(m - 1)$'th symbol in the top row.

Then we can introduce the conditional probability for a certain symbol x given that we have already observed the symbols in the configuration B_m,

$$p(x|B_m) = \frac{p(B_m x)}{p(B_m)} \,. \tag{3.46}$$

This can be interpreted as the conditional probability for the "next" symbol given that we have seen the "previous" $2m(m - 1)$ symbols, i.e., the symbols in the B_m block. The average entropy is

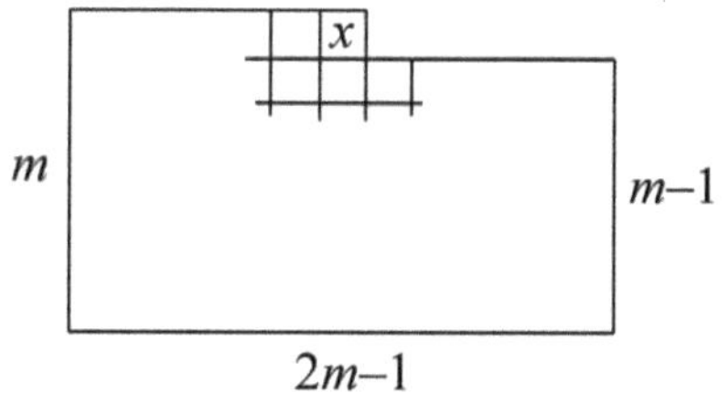

Fig. 3.11 The configuration of cells B_m that is used in the conditional probability for the symbol x in the "next" cell. The cell with x is not part of B_m but is added to form the block $B_m x$. By extending the configuration B_m (increasing m), we get better (more informed) conditional probabilities for the unknown symbol x

$$H_m = \sum_{B_m} p(B_m) \sum_x p(x|B_m) \log \frac{1}{p(x|B_m)} . \tag{3.47}$$

This entropy is analogous to h_m in the one-dimensional case, Eq. (3.12). For $m = 1$, we define $H_1 = S_{1\times 1}$, or the entropy of the single symbol distribution. One can prove, see Helvik and Lindgren (2014), that H_m converges to the entropy, Eq. (3.44), as $m \to \infty$,

$$s = \lim_{m\to\infty} H_m = H_\infty . \tag{3.48}$$

In the same way as the one-dimensional case, the average information of $\log 2$, or equivalently 1 bit, per lattice site can be decomposed into a term quantifying the information in correlations from different lengths (including density information) and a term quantifying the internal randomness of the system,

$$\log 2 = k_{\mathrm{corr}} + s . \tag{3.49}$$

The density information k_1 does not depend on the dimensionality, so it should be as in the one-dimensional case,

$$k_1 = \sum_x p(x) \log \frac{p(x)}{1/2} = \log 2 - S_{1\times 1} = \log 2 - H_1 . \tag{3.50}$$

Then it is clear that if we define correlation information over length m by the difference between two consecutive estimates of the entropy s, i.e., $k_m = -H_m + H_{m-1}$, for $m > 0$, then the decomposition above gives a complete decomposition of the total information per lattice site. In order to show that this definition leads to k_m being a non-negative quantity, we introduce an operator R that reduces a configuration B_m to a configuration $B_{m-1} = RB_m$, by taking away the symbols from the leftmost and rightmost columns as well as from the bottom row. Then k_m can be written as the average relative information when the distribution for "next" symbol given a conditional configuration B_m replaces an a priori distribution with a smaller conditional configuration $B_{m-1} = RB_m$,

$$k_m = -H_m + H_{m-1} = \sum_{B_m} p(B_m) \sum_x p(x|B_m) \log \frac{p(x|B_m)}{p(x|RB_m)} \geq 0 . \qquad (3.51)$$

Here we have used the fact that summation over the symbols in the part that is being reduced in B_m connects the probabilities for B_m and RB_m,

$$p(RB_m) = \sum_{\text{all indices} \notin RB_m} p(B_m) . \qquad (3.52)$$

The correlation information k_{corr} is then decomposed as in the one-dimensional case,

$$k_{\text{corr}} = \sum_{m=1}^{\infty} k_m . \qquad (3.53)$$

This procedure can be repeated in higher dimensions. Note also that the definition of correlation information contains a choice of direction and rotation: the B_m configuration can be chosen in eight different ways. In one dimension there are only two ways to choose, direction left or right, but the choice does not change the resulting quantities. Furthermore, and this holds for both dimension one and higher, one may use other ways to decompose total redundancy which result in different types of correlation measures.

Problems

3.1 Optimal coding

How can the symbol sequences in Example 3.2.3.4 be coded so that one gets a maximally compressed coded sequence?

3.2 Combining processes

Suppose that we have a process that generates completely random binary sequences $(x_0, x_1, x_2, x_3, ...)$ with entropy 1 (bit). Construct a new binary sequence $(y_1, y_2, y_3, ...)$ by addition modulo 2 of pairs of symbols from the first sequence, $y_k = x_k - 1 + x_k \pmod 2$, $k = 1, 2, ...$. What is the entropy of the new sequence?

3.3 FSA example

The finite automaton below represents a stochastic process generating binary sequences in which 1's are separated by an odd number of 0's. What is the Shannon entropy? How long correlations are there (from the information-theoretic point of view)? Assume that the arcs leaving the central node have equal probabilities.

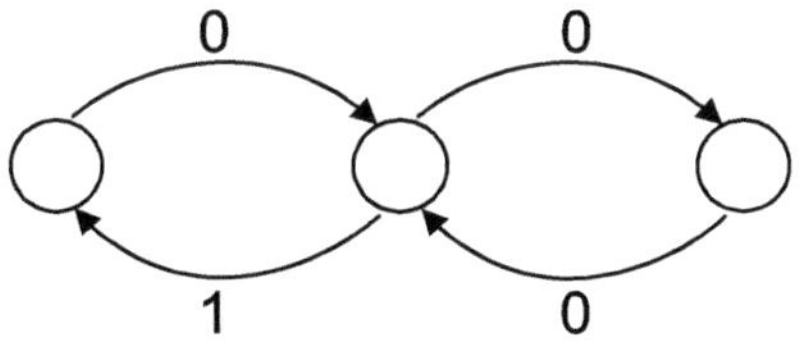

Consider the class of stationary stochastic processes that generate binary symbol sequences in which pairs of 1's are forbidden. What is the largest entropy such a process could generate?

3.4 Information loss in multiplication
Consider two independent stochastic processes generating uncorrelated sequences of symbols 0, 1, and 2, with equal probabilities of the different symbols. Form the product process in which a symbol z_k is the product modulo three of the corresponding symbols from the original processes, $z_k = x_k \cdot y_k \pmod 3$. (Mod 3 means that $2 \cdot 2 \pmod 3 = 1$, while all other multiplications work as usual.) How much information is lost in average from the original pair of symbols (x_k, y_k) when the product z_k is formed?

3.5 Maximise entropy
Consider an infinite sequence of isolated 1's, separated by either one or two 0's (but not more), for example, "...10010101001001010010l...". What probability distribution should we choose for describing the system, if we want the entropy s to be as large as possible? (The answer need not be explicit, but an equation that determines the parameter(s) is sufficient.)

3.6 Two expressions for the entropy of symbol sequences
Show that S_m/m and ΔS_m converges to the same value in the limit of infinite block length, $m \to \infty$, cf. Eq. (3.13), and thus they both quantify the entropy of the symbol sequence.

3.7 Correlation information and complexity
The Hidden Markov model below generates sequences of 0's and 1's. (When two arcs leave a node it is assumed that they have the same probability.)

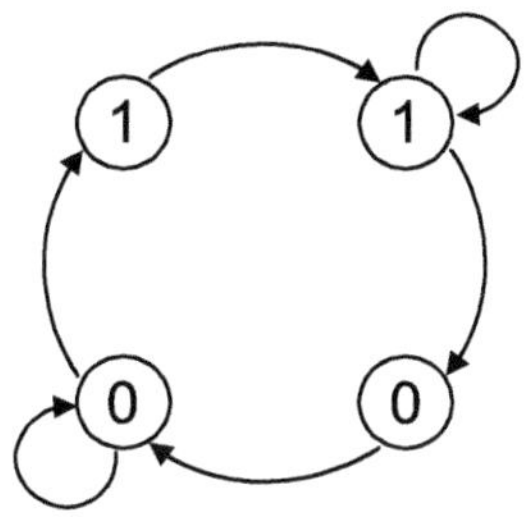

What is the correlation length in information-theoretic terms? Or in other words: What is the longest distance m for which the correlation information is positive, $k_m > 0$. Determine also the correlation complexity η for the system.

3.8 Correlation complexity

Determine the correlation complexity η for the process described by the Hidden Markov model below. (When two arcs leave a node it is assumed that they have the same probability.)

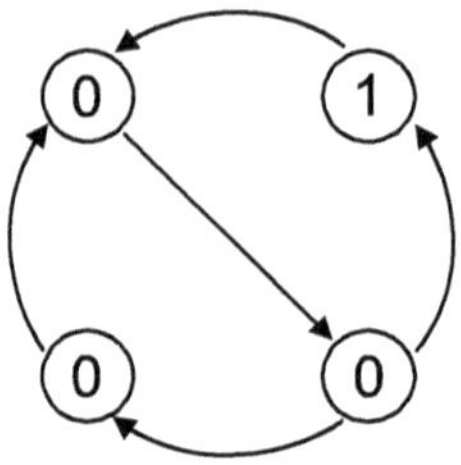

Chapter 4
Cellular Automata

Abstract After a short introduction to cellular automata (CA), an information-theoretic analysis of one-dimensional CA is presented. It is based on a statistical or probabilistic description of the state of the cellular automaton at each time step, where the state is assumed to be an infinite sequence of symbols. Using information theory for symbol sequences, we identify the information contained in correlations, including density information, as well as the remaining randomness, the entropy of the state. We derive laws for development of the entropy over time, and we identify classes of CA dynamics for which these laws take different form: general deterministic, almost reversible, and probabilistic. A methodology for analysing changes in the probabilistic description of the CA state from one time step to the next is described in detail. Illustrations on the evolution of correlation information are given for a number of CA rules.

4.1 Cellular Automata—a Class of Discrete Dynamical Systems

Cellular automata (CA) are simple models of dynamical systems that are discrete in space and time. The number of states per lattice site, or cell, is finite and usually small. The time development is governed by a local updating rule that is applied in parallel over the whole lattice.

John von Neumann in collaboration with Stanislaw Ulam introduced cellular automata in the 1950s, and he wanted to use these models in the study of self-reproduction and noise-sensitivity of computation. One aim was to demonstrate the existence of objects capable of complex behaviour combined with the capability of self-reproduction. This work lead to the design of a cellular automaton rule on a two-dimensional lattice with 29 states per cell. The designed system is capable of making a copy of itself in this space, and it also has the capability to simulate any computational process, usually termed a computationally universal system. In this way complex behaviour of the object was said to be guaranteed. This work was completed and published by Arthur Burks after von Neumann's death (von Neumann and Burks 1966).

© Springer-Verlag GmbH Germany 2024

K. Lindgren, *Information Theory for Complex Systems*, Understanding Complex Systems, https://doi.org/10.1007/978-3-662-68214-2_4

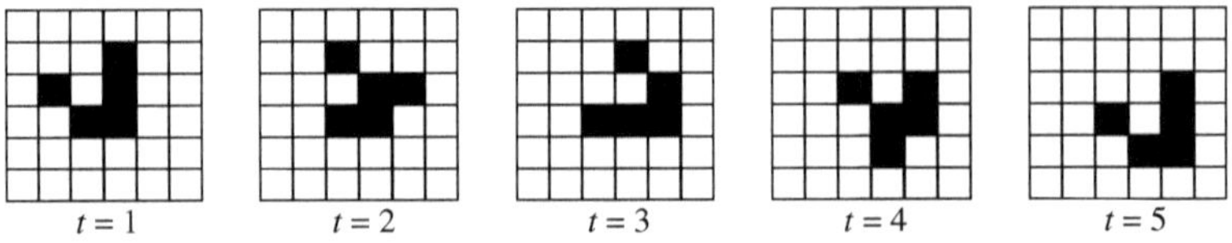

Fig. 4.1 The cellular automaton "Game of Life" exemplified by five time steps of the "glider", a propagating object with an internal cycle of four time steps

There are several possibilities to construct cellular automaton rules, even with only two states per cell, that demonstrate various examples of complex behaviour. One well-known cellular automaton is "Game of Life" that was introduced by John Conway, see Gardner (1970). This rule has been studied extensively, mainly because of its capability of producing complex behaviour like propagating spatio-temporal structures from random initial states, exemplified by the simple "glider" in Fig. 4.1. The rule is based on a local configuration involving the cell and its 8 neighbours. Cells can be "alive" or "dead", represented by black and white in the figure. A "dead" cell becomes "alive" in the next time step if exactly 3 neighbours are "alive", while a cell that is "alive" remains so only if 2 or 3 neighbours are "alive".

One can show that "Game of Life" also has the capability of universal computation. This can be done, for example, by constructing structures in the lattice that function as wires that allow for propagating signals that may interact through structures serving as logical gates.

Already in the simplest class of CA, the CA rules in one dimension that have local interaction depending on nearest neighbours only, there are examples of various types of complex behaviour. In 1989 it was shown that with 7 states per cell, it is possible to construct CA rules that are capable of universal computation (Lindgren and Nordahl 1990). A direct implication is that results from computation theory applies to this rule. For example, the halting problem, stating that there is no general procedure to determine whether a computer program will ever halt, transforms into a theorem for computationally universal CA. This means that there are initial states for such a CA for which it is impossible to prove whether the CA will develop to a fixed point. In fact, it has been shown that the even more simple CA rule R110, depending on only two states per cell, is computationally universal (Cook 2004; Wolfram 2002). The space-time pattern of R110 is exemplified in Fig. 4.2.

Cellular automata as models for physical systems met with a renewed interest in the 1980s, partly because of the classic paper by Stephen Wolfram (1983), "Statistical mechanics of cellular automata." Among the physical applications of cellular automata, the "lattice gases" are the most well known. These CA simulate systems of particles with discrete (usually unit) velocities moving on a lattice. Despite the highly simplified microscopic dynamics, some of these systems approximate the Navier-Stokes equations for fluid dynamics, when averages are taken over large numbers of particles.

In this chapter we shall demonstrate how the information-theoretic concepts presented in Chap. 3 can be used in order to analyse how order and disorder develop

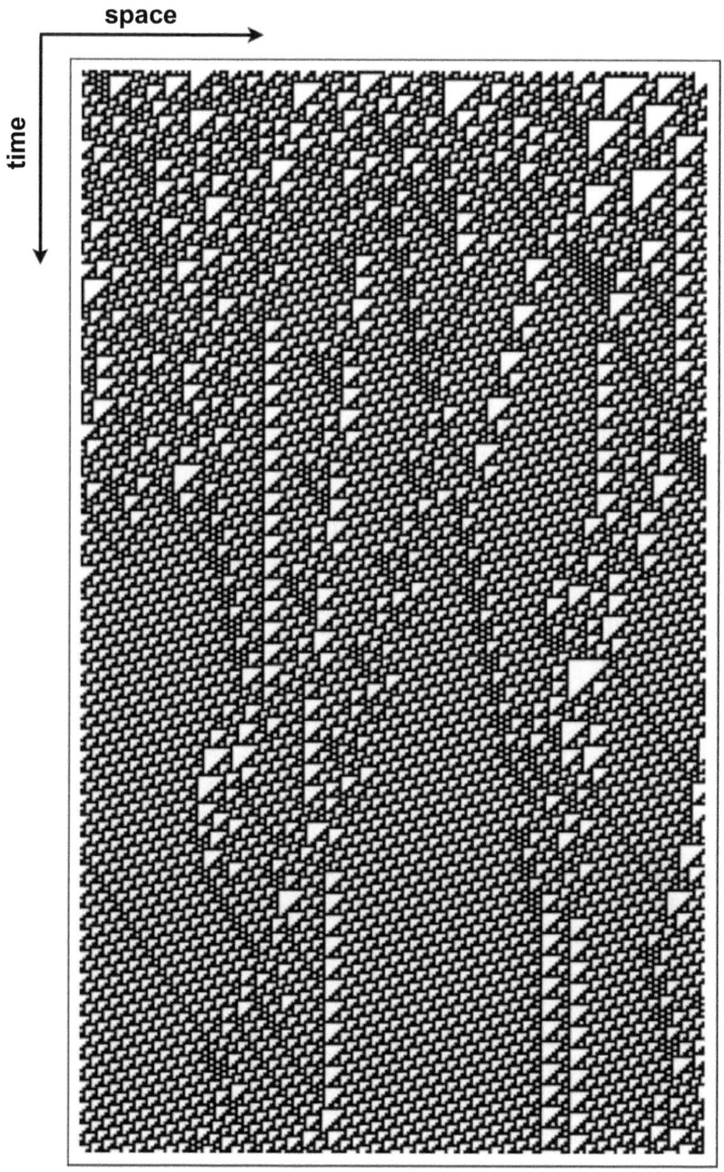

Fig. 4.2 The time evolution of CA rule R110 starting with a "random" initial state (in the top row) shows how a periodic background pattern is built up at the same time as complex structures propagate and interact. This is a good example of the complexity that simple CA rules may exhibit

during the time evolution for different types of cellular automaton rules. We shall in particular study CA that are reversible and discuss how apparently random patterns may be created in such systems. This is related to the second law of thermodynamics to be discussed in a later chapter.

4.2 Elementary Cellular Automata

The simplest class of cellular automata is based on a one-dimensional lattice with two states per cell and a nearest neighbour rule for the dynamics. Such a rule is fully determined by specifying the next state of a cell for each of the eight possible local states that describe the present state of the local neighbourhood. An example of such a specification is shown in Table 4.1. The time evolution of this rule, starting from a random sequence of 0's and 1's, is exemplified in Fig. 4.2. The top row of black (1) and white (0) pixels represents the initial state, and the following rows represent the development in time when the rule in the table is applied in parallel over the whole row. In this CA class the rules can thus be uniquely described by the binary digits in the second row of the table, which means that eight binary symbols determine the rule. In the example, we have the rule number $(01101110)_2$ which in decimal form is 110—the rule mentioned above being computationally universal. There are $2^8 = 256$ elementary CA rules, but several of these are equivalent (by symmetries, such as changing symbols and direction), which results in 88 different rules.

The dynamic behaviour of elementary CA can differ a lot from one rule to another. Wolfram suggested a classification with four types. The simplest CA, class I, approach a homogenous fixed point as is exemplified by the rule in Fig. 4.3a. A class II rule develops into an inhomogeneous fixed point or to a periodic and/or simple shift of the pattern like in Fig. 4.3b. The class III rules never seem to approach an ordered state, but their space-time patterns continue to look disordered, as illustrated in Fig. 4.3c. These rules are often called "chaotic". Then there is a class IV that is more vaguely defined as a border class between II and III with long complex transients, possibly mixed with a spatio-temporal periodic background pattern, see Fig. 4.3d. Also rule R110 in Fig. 4.2 belongs to this class.

An important characteristic of the class III, or the chaotic, rules is that they are sensitive to small perturbations, just like chaotic low-dimensional systems. If one follows the time evolution starting from two initial states, differing at one position only, the number of differing positions tends to increase linearly in time. This is

Table 4.1 Cellular automaton rule

t	111	110	101	100	011	010	001	000
$t+1$	0	1	1	0	1	1	1	0

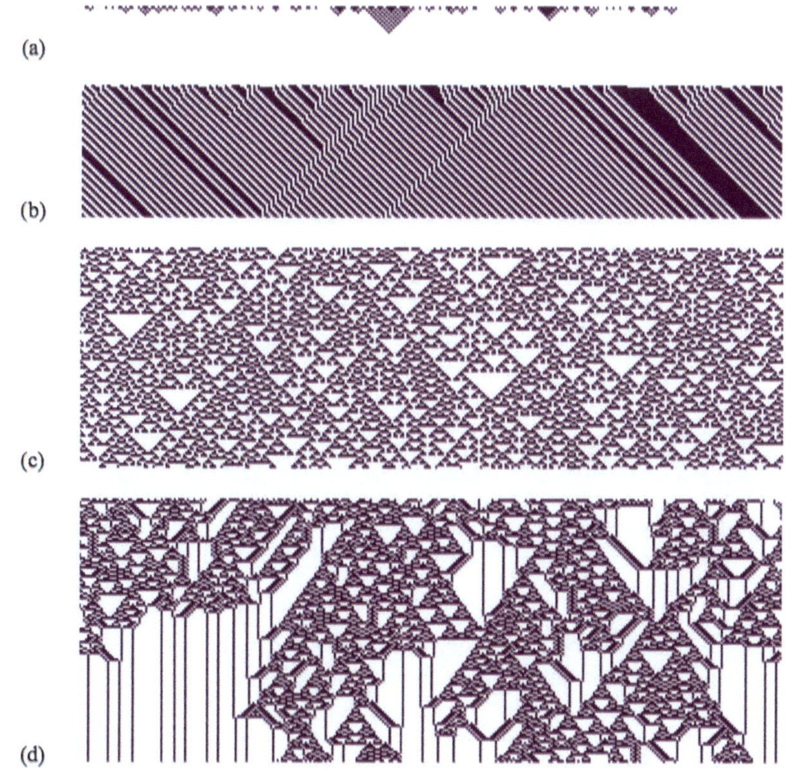

Fig. 4.3 Cellular automata space-time patterns from the four classes of CA rules: **a** Class I rules approach a fixed point (e.g., R160). **b** Class II rules develop a periodic pattern in space and/or time (e.g., R213). **c** Chaotic or class III rules are characterised by a continuous change of the patterns at the same time as a high disorder is kept (e.g., R22). **d** Class IV is a border class involving features from both class II and class III with long complex transients. The last example is generated by a rule that depends on the number of living cells (1's) in a 5-cell neighbourhood, so that a cell survives if 1 or 3 neighbours are alive and a cell becomes living if 2–4 neighbours are alive; otherwise the cell dies

illustrated in Fig. 4.4, which shows the differing cells between the space-time patterns of two CA following rule R22.

4.3 Information Theory for Cellular Automata

What happens with the entropy that describes the internal disorder of the CA state in the time evolution? Initially we may have prepared the system as a maximally disordered sequence of symbols with an entropy of 1 bit per cell. Under what circumstances will this entropy decrease to create order in the system?

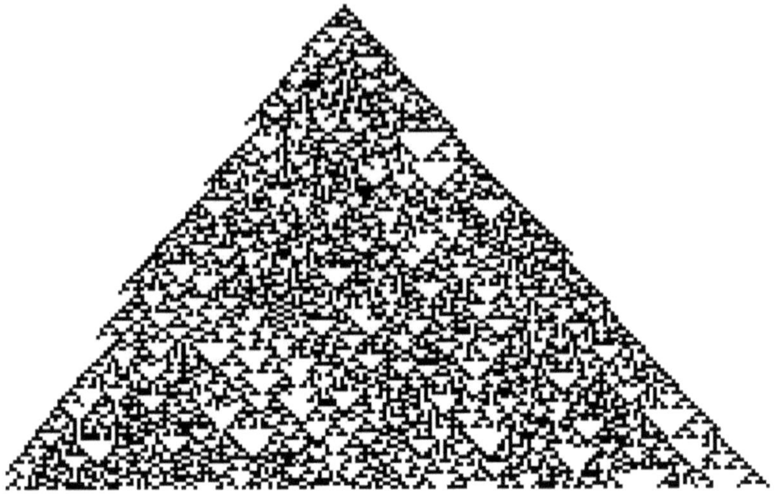

Fig. 4.4 The difference pattern resulting from the time evolution of CA rule R22 starting from two different states, initially differing only by the state in one cell

We shall use the information theory for symbol sequences presented in Chap. 3, in order to prove relations between the entropy of the CA at time t and the following time step $t + 1$, see Lindgren (1987). Suppose that we have a cellular automaton rule with range r, i.e., the local neighbourhood involved in the rule includes $2r + 1$ cells (r to the left, r to the right, and the cell itself). We also assume that there are only two states per cell (0 and 1); the formalism is easily extended to more states. At each time step t, the state of the CA is described as a symbol sequence characterised by its entropy per symbol $s(t)$, as a measure of its internal randomness given by Eq. (3.9).

4.3.1 Deterministic Rules

To begin with we shall assume that the rules are deterministic, which implies that the state at time t fully determines the state at the next time step. Let β_m denote a certain sub-sequence of symbols of length m at time $t + 1$. This sequence may have several possible predecessors at time t. Since the rule has range r, the predecessor sequence has length $m + 2r$, and we denote such a sequence α_{m+2r}. The probability distributions that describe the symbol sequences at different times may of course be different, and we use p for probabilities at time t and p' for probabilities at time $t + 1$. Then the probability for a sequence β_m at $t + 1$ can be expressed as the sum of the probabilities for all possible predecessors,

$$p'(\beta_m) = \sum_{\alpha_{m+2r} \to \beta_m} p(\alpha_{m+2r}) \ . \tag{4.1}$$

This relation can be used to establish a connection between the block entropy, see Eq. (3.8), of length $m + 2r$ at time t with the block entropy of length m at time $t + 1$,

$$S_m(t+1) = \sum_{\beta_m} p'(\beta_m) \log \frac{1}{p'(\beta_m)} =$$
$$= \sum_{\beta_m} \left(\sum_{\alpha_{m+2r} \to \beta_m} p(\alpha_{m+2r}) \right) \log \frac{1}{\sum_{\alpha_{m+2r} \to \beta_m} p(\alpha_{m+2r})} \leq$$
$$\leq \sum_{\beta_m} \sum_{\alpha_{m+2r} \to \beta_m} p(\alpha_{m+2r}) \log \frac{1}{p(\alpha_{m+2r})} = S_{m+2r}(t) \,. \tag{4.2}$$

Here we have made use of the fact that $\log(1/x)$ is a decreasing function of x, when removing the summation in the logarithm.[1] This inequality can be used to derive the change in entropy (per symbol), Eq. (3.9), between two successive time steps in the CA time evolution,

$$\Delta_t s(t) = s(t+1) - s(t) = \lim_{m \to \infty} \left(\frac{1}{m} S_m(t+1) - \frac{1}{m+2r} S_{m+2r}(t) \right) =$$
$$= \lim_{m \to \infty} \left(\frac{1}{m} (S_m(t+1) - S_{m+2r}(t)) + \left(\frac{1}{m} - \frac{1}{m+2r} \right) S_{m+2r}(t) \right) \,. \tag{4.3}$$

By using Eq. (4.2) and the fact that the second term in Eq. (4.3) goes like $2rs(t)/m$, and approaches 0 as $m \to \infty$, we can conclude that, for deterministic cellular automaton rules, the entropy decreases (or stays constant) in the time evolution,

$$\Delta_t s(t) \leq 0 \,. \tag{4.4}$$

This irreversibility, when the inequality is strict, can be understood by the fact that deterministic CA rules reduce the number of possible states (symbol sequences) in the time evolution. This decrease in entropy is associated with an increase in total correlation information (including density information). This is clearly visible in the periodic pattern formed in the time evolution of rule R110, illustrated in Fig. 4.2. The initial state is a completely disordered sequence of zeroes and ones with en entropy $s = 1$ (bit), but as time goes on correlation information is built up and entropy decreases.

[1] We use the fact that $(x+y)\log(1/(x+y)) \leq x \log(1/x) + y \log(1/y)$.

4.3.2 Almost Reversible Rules

An important property of the rules that govern a dynamical system is whether they can be considered reversible. In physical systems, the microscopic laws of motion usually are reversible, and therefore it is of interest to investigate the information-theoretic properties of cellular automata that can be considered reversible. Intuitively, we should expect the entropy of such a CA to be constant in time.

With a *reversible* CA rule we consider a CA in which each microstate (infinite symbol sequence) has a unique predecessor at the previous time step. We shall show, though, that there is a weaker form of reversibility that results in a constant entropy, i.e., equality in Eq. (4.4).

Suppose that a cellular automaton rule R with range r can be written as

$$R(x_1, x_2, ..., x_{2r}, x_{2r+1}) = f(x_1, x_2, ..., x_{2r}) + x_{2r+1} \bmod 2 \,, \qquad (4.5)$$

where f is some binary-valued function, and the summation is taken modulo 2. This means that the rule R is a one-to-one mapping between its last argument, x_{2r+1}, and the result—a flip of the state in the rightmost cell in the neighbourhood flips the result of the rule. (A corresponding type of rule can be constructed with respect to the leftmost argument.)

Note that this type of rule involves a certain type of reversibility: Suppose that we know the semi-infinite microstate $y_1y_2y_3...$ of the CA at time $t+1$, and that we know the first $2r$ cells of the corresponding preceding microstate $x_1x_2...x_{2r}$ at time t, which is then all but one of the cells in the neighbourhood that determines y_1. Then we can use the rule, Eq. (4.5), by entering $x_1x_2...x_{2r}$ as an argument in f, to find the cell state x_{2r+1} at t that corresponds to the result y_1. This means that we now have all but the rightmost cell in the neighbourhood that determines y_2, and the same procedure can be repeated again to derive x_{2r+2} from y_2. By repeated use of this procedure, we may reproduce the complete semi-infinite microstate $x_1x_2x_3...$ at time t. Only $2r$ bits of information are needed (the first $2r$ cells) to reproduce the preceding microstate from the present one, and we call such a rule *almost reversible*. These rules differ from the reversible rules in that each microstate may have several (up to 2^{2r}) preceding microstates.

The probability for a sequence β_m at $t+1$ can be expressed as the sum of the probabilities for all possible ancestors using a transfer function $T(\alpha_{m+2r}, \beta_m)$. The function T takes the value 1 if $\alpha_{m+2r} \rightarrow \beta_m$ under the rule, otherwise 0. We can express the probabilities p' at time $t+1$ in terms of the probabilities p from the previous time step,

$$p'(\beta_m) = \sum_{\alpha_{m+2r}} T(\alpha_{m+2r}, \beta_m)p(\alpha_{m+2r}) \,. \qquad (4.6)$$

The transfer function has the properties

$$\sum_{\alpha_{m+2r}} T(\alpha_{m+2r}, \beta_m) = 2^{2r} , \tag{4.7}$$

$$\sum_{\beta_m} T(\alpha_{m+2r}, \beta_m) = 1 . \tag{4.8}$$

To simplify the notation, we drop the length indices on α and β. The difference in block entropies between m-length blocks at $t+1$ and $(m+2r)$-length blocks at t can then be written

$$\begin{aligned} S_m(t+1) - S_{m+2r}(t) &= \sum_\beta p'(\beta) \log \frac{1}{p'(\beta)} - \sum_\alpha p(\alpha) \log \frac{1}{p(\alpha)} = \\ &= \sum_\beta \sum_\alpha T(\alpha, \beta) p(\alpha) \log \frac{p(\alpha)}{p'(\beta)} = \\ &= \sum_\beta p'(\beta) \sum_\alpha \frac{T(\alpha, \beta) p(\alpha)}{p'(\beta)} \log \frac{T(\alpha, \beta) p(\alpha) / p'(\beta)}{T(\alpha, \beta) / \sum_{\alpha'} T(\alpha', \beta)} - \log \sum_{\alpha'} T(\alpha', \beta) . \end{aligned} \tag{4.9}$$

Here we see that the sum over α in the last row is a relative information, since both $T(\alpha, \beta) p(\alpha) / p'(\beta)$ and $T(\alpha, \beta) / \sum_{\alpha'} T(\alpha', \beta)$ are normalised probability distributions over α whenever $p'(\beta) > 0$. Since the relative information is non-negative and the last logarithm is $2r$, we can conclude that

$$S_m(t+1) - S_{m+2r}(t) \geq -2r . \tag{4.10}$$

Using the same approach as in Eq. (4.3) we conclude that

$$\begin{aligned} \Delta_t s(t) &= \lim_{m \to \infty} \left(\frac{1}{m} S_m(t+1) - \frac{1}{m+2r} S_{m+2r}(t) \right) \geq \\ &\geq \lim_{m \to \infty} \left(\frac{2r}{m(m+2r)} S_{m+2r}(t) - \frac{2r}{m} \right) = 0 . \end{aligned} \tag{4.11}$$

In combination with the law of non-increasing entropy for deterministic rules, Eq. (4.4), this results in a constant entropy in the time evolution of almost reversible rules,

$$\Delta_t s(t) = 0 . \tag{4.12}$$

For these cellular automata the initial distribution of information between entropy and correlation information (redundancy) is kept in the time evolution. Still, there may be non-trivial information-theoretic changes since the correlation length may change, provided that the system is prepared with an initial correlation information, for example in the form of density information. We will look at illustrations of this in Sect. 4.4.

4.3.3 Rules with Noise

If noise, in the form of a process that randomly selects cells in which the state is changed, interferes with an otherwise deterministic rule, it seems reasonable that an increase in entropy is possible for some rules, provided that we start from an initial state with some redundancy, $s < 1$. Suppose that the deterministic rule R is applied as usual on the complete CA microstate, but that in the resulting microstate each cell state is flipped with a probability q. Then we denote such a probabilistic rule by the pair (R, q).

The change in entropy in one time step, can then be decomposed into two terms, one negative (or zero) term $\Delta_R s(t)$ associated with the deterministic change resulting from rule R, cf. Eq. (4.4), and one term $\Delta_q s(t)$ due to the entropy change from the noise,

$$\Delta_t s(t) = \Delta_R s(t) + \Delta_q s(t) \ . \tag{4.13}$$

Let us now investigate how the noise affects the entropy s. Suppose that the probability distribution for m-length sequences of symbols are given by probabilities $p(\alpha_m)$, after the rule R has been applied. The noise then transforms the probabilities according to a transfer function $T_q(\alpha_m, \beta_m)$,

$$\hat{p}(\beta_m) = \sum_{\alpha_m} T_q(\alpha_m, \beta_m) p(\alpha_m) \ . \tag{4.14}$$

The transfer function, which is simply the probability that a certain sequence α_m is transformed into β_m, depends on the Hamming distance $H(\alpha_m, \beta_m)$, i.e., the number of cells for which the disturbed sequence β_m differs from the original one α_m. This implies that

$$T_q(\alpha_m, \beta_m) = q^{H(\alpha_m, \beta_m)} (1 - q)^{m - H(\alpha_m, \beta_m)} \ . \tag{4.15}$$

Normalization and symmetry implies that

$$\sum_{\alpha_m} T_q(\alpha_m, \beta_m) = \sum_{\beta_m} T_q(\alpha_m, \beta_m) = 1 \ . \tag{4.16}$$

The change in block entropy $\Delta_q S_m$ due to the noise can then be written (dropping the length notation m on the sequence variables α and β)

$$
\begin{aligned}
\Delta_q S_m &= S[\hat{P}_m] - S[P_m] = \\
&= \sum_\beta \sum_\alpha T_q(\alpha,\beta) p(\alpha) \log \frac{1}{\hat{p}(\beta)} - \sum_\alpha p(\alpha) \log \frac{1}{p(\alpha)} = \\
&= \sum_\beta \sum_\alpha T_q(\alpha,\beta) p(\alpha) \log \frac{T_q(\alpha,\beta) p(\alpha)/\hat{p}(\beta)}{T_q(\alpha,\beta)} = \\
&= \sum_\beta \hat{p}(\beta) \sum_\alpha \frac{T_q(\alpha,\beta) p(\alpha)}{\hat{p}(\beta)} \log \frac{T_q(\alpha,\beta) p(\alpha)/\hat{p}(\beta)}{T_q(\alpha,\beta)} \geq 0\,. \qquad (4.17)
\end{aligned}
$$

Here we have used the fact that the sum over α in the last line is the relative information (for each sequence β) between an "a priori" distribution $T_q(\alpha,\beta)$ and a "new" distribution $T_q(\alpha,\beta)p(\alpha)/\hat{p}(\beta)$, which then guarantees that the change in block entropy is non-negative. From this we can conclude that

$$
\Delta_q s(t) = \lim_{m\to\infty} \frac{1}{m} \Delta_q S_m \geq 0\,. \qquad (4.18)
$$

But it is also clear, from the fact that the relative information is zero only when both involved distributions are identical, that the added noise entropy is zero only when $p(\alpha_m) = \hat{p}(\beta_m)$ for all α_m and β_m. This only occurs when $p(\alpha_m) = \hat{p}(\beta_m) = 2^{-m}$, i.e., when the cellular automaton state is completely disordered with maximum entropy, $s = 1$.

We can now conclude that if the CA rule R is almost reversible, the only entropy change comes from the added noise. Starting from an ordered initial condition, $s < 1$, the entropy will increase until the system is completely disordered,

$$
\Delta_t s(t) > 0\,, \text{ if } s(t) < 1\,, \qquad (4.19)
$$

$$
\Delta_t s(t) = 0\,, \text{ if } s(t) = 1\,. \qquad (4.20)
$$

The results on the entropy change in deterministic and noisy cellular automata are summarised in Table 4.2. For irreversible and noisy CA both increase and decrease in entropy is possible, depending on rule and the details of the current microstate.

The results presented in this section are easily extended to two and more dimensions. Several CA with physical relevance are of interest for this type of analysis like

Table 4.2 Cellular automata entropy laws

	Deterministic	Noisy
Irreversible	$\Delta_t s(t) \leq 0$	–
Rreversible	$\Delta_t s(t) = 0$	$\Delta_t s(t) \geq 0$

lattice gas models and models of spin system dynamics. The latter will be dealt with in the next chapter. As we will see, the entropy associated with the symbol sequence of a microstate is proportional to the thermodynamic entropy, and therefore the different relations shown in Table 4.2, especially for (almost) reversible rules, have implications for the second law of thermodynamics, stating that the thermodynamic entropy for a closed system cannot decrease.

4.4 Examples of Information-Theoretic Properties in the Evolution of Simple CA

The irreversibility expressed by the law of non-increasing entropy for deterministic rules, Eq. (4.4), and illustrated by the time evolution of rule R110 in Fig. 4.2, can be quantified by calculating the increase in correlation information during the first time steps. This is depicted in Fig. 4.5 showing the contributions to the redundant information from density information k_1, as well as from correlation information from blocks of lengths up to 8, i.e., k_2, ..., k_8, as functions of time. The initial state is completely "random" with an entropy $s = 1$ and zero correlation information, $k_{\mathrm{corr}} = 0$. During the time evolution the entropy is transformed to density and correlation information.

If an almost reversible rule starts from a random initial state ($s = 1$) bit, then the maximum entropy will be conserved in the time evolution, according to Eq. (4.12).

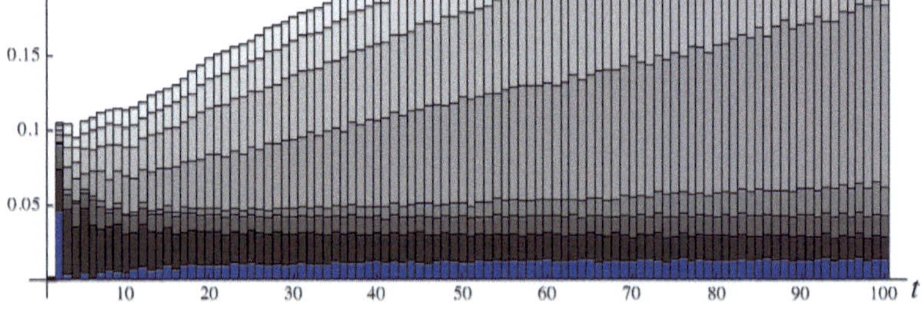

Fig. 4.5 During the time evolution of CA rule R110, density and correlation information increases at the same time as entropy decreases by the same amount. In the diagram, the density information k_1 is represented by blue and contributions to correlation information from the first 7 terms, k_2 to k_8, represented by grayscale patterns from dark to light gray, are put on top of each other

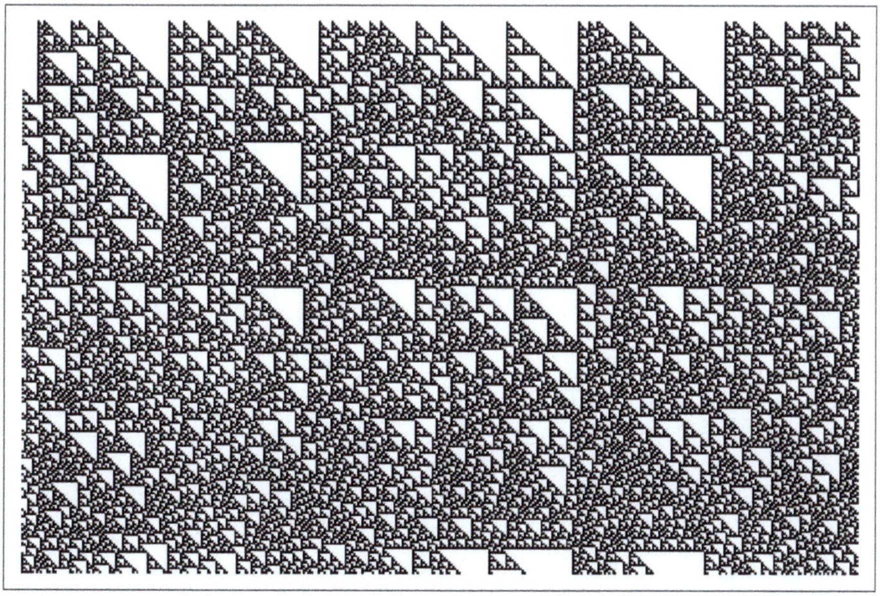

Fig. 4.6 The space-time pattern of rule R60, an almost reversible rule, exhibits non-trivial behaviour when there is redundancy present in the system, $s < 1$ (bit). The initial state is dominated (90%) of 0's

For this type of system it is more interesting to study what is happening if one starts with a low entropy initial state, i.e., an initial configuration with $s < 1$ bit. In the example shown in Fig. 4.6, the time evolution of rule R60 is shown. This is an almost reversible rule given by $x'_k = R(x_{k-1}, x_k) = x_{k-1} + x_k \pmod 2$. This type of rule is also called an *additive rule*. Note that the rule only depends on the cell itself and the left neighbouring cell. The initial state is prepared with a 90% frequency of 1's, but without any correlations, resulting in $k_1 \approx 0.53$ bit and $s \approx 0.47$ bit.

The time evolution of the CA in Fig. 4.6 is characterised by the density information and the short length correlation information terms, see Fig. 4.7. Even though the total correlation information is conserved, there are large changes in the specific k_m terms, showing that correlation information is changing significantly in length from one time step to another. Note also that at certain time steps, $t^* = 1, 2, 4, 8, 16$, etc., a large part of the correlation information is again gathered in the density information, k_0, and that these time steps are preceded by a series of steps in which correlation information is moving back from larger distances towards the density (or single cell) information term. Contributions from correlation information of lengths larger than 8 is not shown in the figure, but since we know that the total correlation information is conserved we also know that these contributions add up to the horizontal line at $k = 1 - s$ (bits). Of course, as time goes on, these events become more rare, which implies that in the long run most of the states (i.e., most of the time steps) will not display any short length correlations. One can actually quantify how the correlation

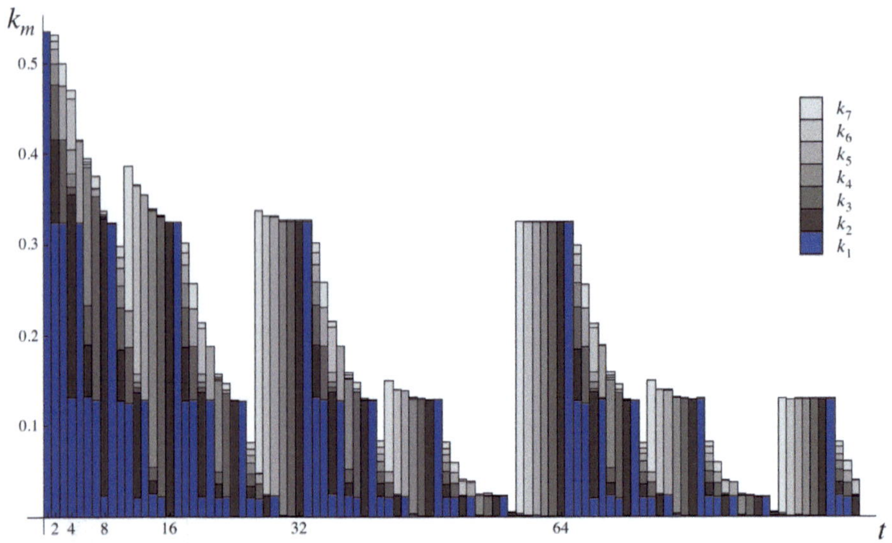

Fig. 4.7 The diagram shows the contributions to the correlation information from density information (blue) and correlation information of lengths 2 up to 7 (grayscale from dark to light) put on top of each other, for the time evolution of the almost reversible rule R60. The initial state does not contain any correlations but there is density information due to the 90% dominance of the state 0. Therefore, the initial redundancy is larger than zero and the entropy $s < 1$ (bit). The reversibility implies that the entropy is conserved, and the figure then shows that information quickly is distributed over correlations of lengths larger than 7

information becomes more important on longer distances by using the correlation complexity η, Eq. (3.36). It can be shown (Lindgren and Nordahl 1988) that for rule R60 the complexity increases linearly if $s < 1$ (bit),

$$\eta(t) = t \cdot s \ . \tag{4.21}$$

If the entropy is $s = 1$ (bit), though, the CA state remains fully random, and the correlation complexity is constantly 0. Thus, Since $\eta = k_{\text{corr}} l_{\text{corr}}$, Eq. (3.37), the average correlation length l_{corr} increases linearly in time, provided that $s < 1$ (bit).

Therefore, even if the entropy is conserved for rule R60, the state of the CA will in the long run (at most time steps) appear more and more disordered, unless increasingly long correlations are taken into account when calculating the entropy. The low initial entropy may appear to increase, if the system is observed locally only.

If the time evolution of rule R60 is disturbed by random noise, the correlations that are built up from the initial density information are rapidly destroyed. In Fig. 4.8, the space-time pattern is shown when a noise of 1% ($q = 0.01$) is added. It is clear, as also Fig. 4.9 illustrates, that the time steps when a large part of the correlation information is recollected in the density information disappear. According to Eqs. (4.19)–(4.20),

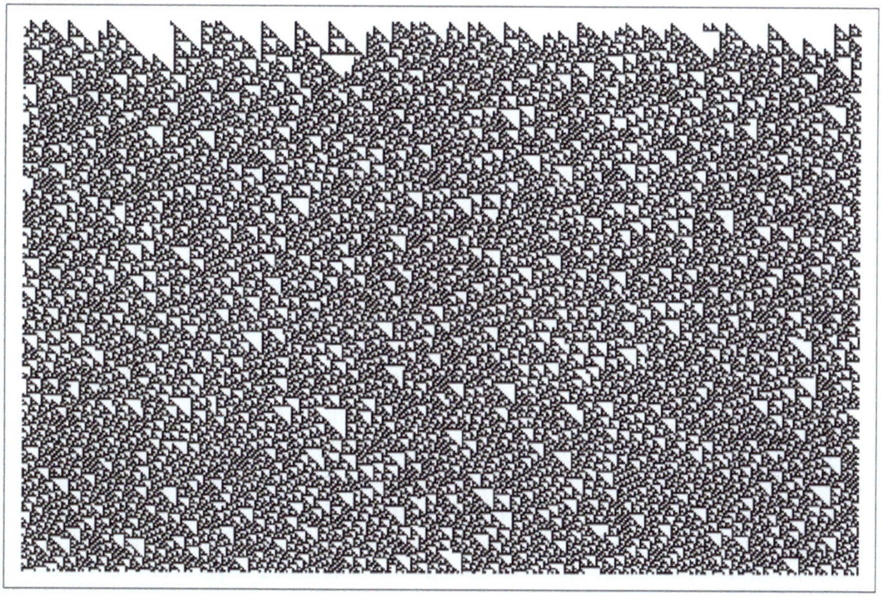

Fig. 4.8 The space-time pattern for rule R60, where noise at a level of 1% is added, is shown. The initial redundancy, based on the 90% dominance of state 0, is efficiently destroyed in the time evolution, cf. Fig. 4.6

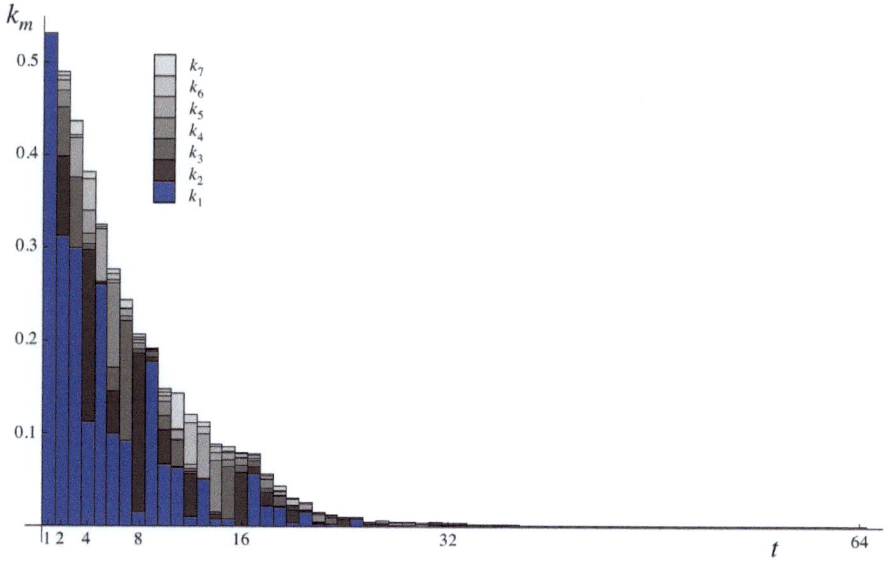

Fig. 4.9 In the same way as was illustrated for rule R60 without noise, see Fig. 4.7, the contributions to the correlation information, from density information (blue) and from correlation information over lengths 2–7 (dark to light grey), are put on top of each other. It is clear that the noise destroys the correlations, and the entropy increases

the time evolution will lead to a completely disordered state with maximum entropy $s = 1$ (bit).

4.5 Analysis of CA Time Evolution Using Hidden Markov Models

The change of information-theoretic characteristics in the state of a one-dimensional CA, represented by the infinite sequence of symbols, from one time step to the next, can be analysed by investigating how the finite state automaton (FSA) representation of the sequence changes under the CA rule. Here we will assume that the state is given by a stochastic process, represented by a certain FSA. The FSA can correspond to a Markov process, which it does for the initial state if we, for example, start with uncorrelated symbols, like in the examples in previous sections. In general, though, also for such an initial condition, the CA rule transforms the automaton to one representing a hidden Markov model, as we shall see in examples below. How the procedure works to transform one FSA to a new one under a CA rule will be illustrated by the following example.

Assume that we have, at time t, a certain description of the CA state in the form of the hidden Markov model shown in Fig. 4.10a, where we assume all transition probabilities to be 1/2 whenever there is a choice. First we rewrite that into a corresponding Markov model, Fig. 4.10b, in which all the states are represented by specific symbols.

Next we construct a new Markov model out of the one in Fig. 4.10b, by having each new FSA state representing pairs of symbols from the sequence. These pairs are overlapping, so that a state can be seen as a two-symbol wide "window, and each transition in this automaton moves the window one step forward in the sequence of symbols, as illustrated in Fig. 4.11.

When we have the FSA in this form, still representing the CA state at time t, we can use the CA rule to determine how the FSA will be transformed to a new one representing the stochastic process that describes the CA state at time $t + 1$.

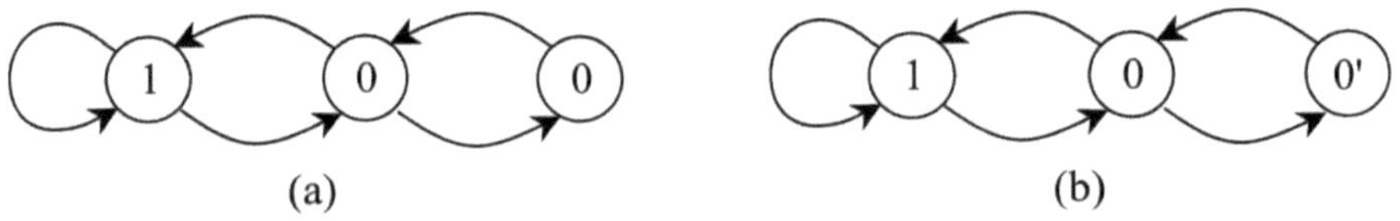

Fig. 4.10 **a** FSA representing the hidden Markov model in which sequences of symbol 1 are separated by an odd number of the symbol 0. **b** FSA representing the corresponding Markov model in which the rightmost state has been given symbol 0' in order to distinguish between the two 0-states. The hidden Markov model in **a** is then given by the function f(1) = 1, f(0) = 0, and f(0') = 0. All transition probabilities are 1/2 when there is a choice

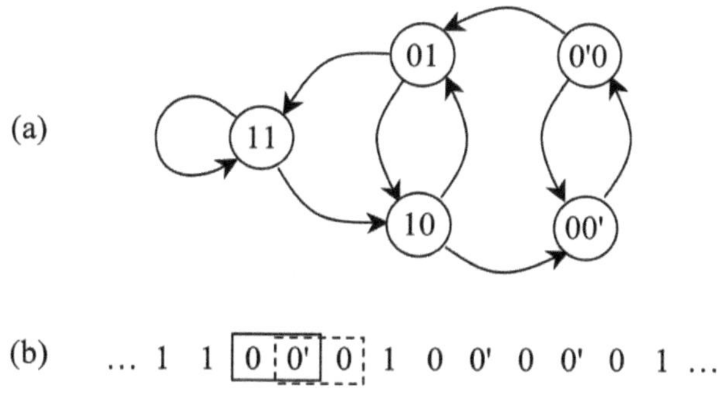

Fig. 4.11 (**a**) FSA representing the same Markov model as in Fig. 4.10b, but where states correspond to overlapping pairs of symbols, and where the transition move to the next pair overlapping with the previous one, as illustrated in (**b**) The transition probabilities from the original automaton remains, all being 1/2 when there is a choice

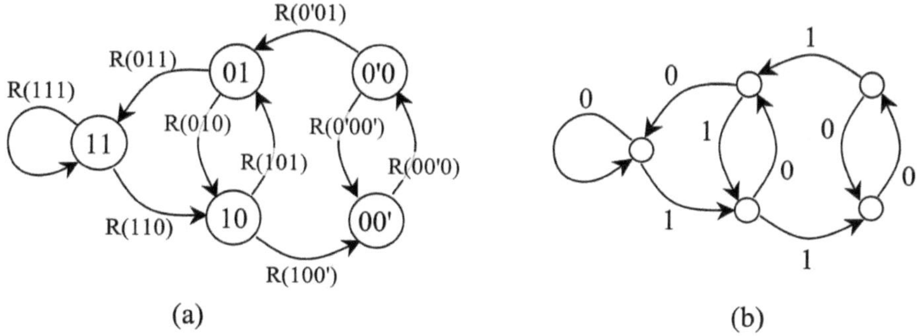

Fig. 4.12 (**a**)The transitions on the FSA for the original time step t correspond to triplets of symbols and thus defines the local neighbourhood on which the CA rule is applied. In (**b**) the resulting FSA is shown where the states in the nodes are hidden, and this FSA represents the stochastic process describing the CA state at time $t+1$

The transitions in the FSA based on pairs of symbols, in Fig. 4.11, correspond to triplets of symbols (resulting from the two overlapping pairs of symbols of the states connected by the transition). We can apply the elementary CA rule on the triplets to get the automaton representing the FSA at time $t+1$. In Fig. 4.12 this is illustrated with the resulting automaton shown for rule R86, see Table 4.3. This rule is almost reversible (involving a 1-to-1 mapping between the rightmost cell and resulting state). Thus we know that the entropy s will not change under the CA rule, but correlation characteristics may do.

It is now possible to use the FSA of Fig. 4.12b to calculate the properties of the CA at the new time step. One can simplify these calculations, though, by reducing the size of the FSA exploiting the fact there there exist equivalent states in the automaton. The leftmost state (old `11`) and the top middle state (old `01`) both produce a `0` with probability $1/2$ ending up in the leftmost state and they both produce a `1` with the

Table 4.3 CA rule 86

t	111	110	101	100	011	010	001	000
$t+1$	0	1	0	1	0	1	1	0

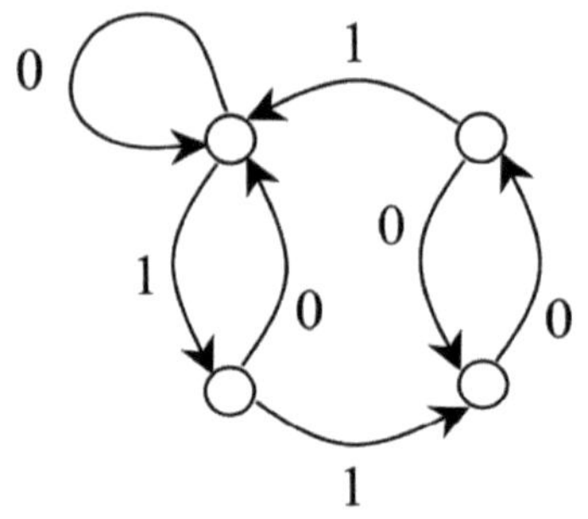

Fig. 4.13 The most compact FSA representation of the CA state characteristics for time $t+1$, when elementary CA rule R86 is applied to a CA state characterised by the FSA in Fig. 4.10 at time t. (It is straightforward to transform this FSA automaton to an equivalent one with the states in the nodes instead of on the arcs)

same probability ending up in the bottom middle state (old `10`). Therefore they are equivalent in every aspect, and the states can be merged as shown in Fig. 4.13.

For the FSA in Fig. 4.13, it is straightforward to calculate the stationary distribution over the nodes, which results in probability 2/5 for the upper left node and 1/5 for the others. One can check that the conditions for using the simplified expression of the entropy, as discussed in Sect. 3.2.2, is fulfilled here.[2] This means that the entropy equals the information required for the choices of transitions in the FSA. Each such choice involves 1 bit, as the probabilities are 1/2 whenever there are two outgoing transitions. This occurs in all nodes except the lower right one, with a total weight of 4/5. This results in the entropy $s = 4/5$ (bits), which one also finds for the original FSA in Fig. 4.10a.

The correlation complexity η for the FSA at $t+1$ in Fig. 4.13 can be calculated using the approach in Sect. 3.3.1 and Eq. (3.42), but not Eq. (3.43) since a future sequence does not uniquely determine the start node. Numerically one can find that $\eta_{t+1} \approx 0.9613$. For the original FSA at t the correlation complexity is similarly derived analytically: $\eta_t = \log 5 - (8/5)\log 2 \approx 0.7219$ (bits). This indicates an increase in average correlation length for this time step.

[2] For almost all preceding sequences (generated by this FSA), in the infinite length limit, one can uniquely determine which node is the final one.

4.6 Local Information Detecting Patterns in CA Time Evolution

In the time evolution of a cellular automaton rule it is often difficult to distinguish the normal behaviour from more rare events. For example, in a chaotic rule like R18 the irregular space-time pattern contains local structures that are less common, but they are not easy to distinguish. Therefore it would be useful to be able to filter out the regular patterns to identify local configurations that deviate. Below follows a brief introduction to the local information framework developed in Helvik et al. (2004, 2007). Similar information filters have been developed by Lizier and coworkers, see, e.g., Lizier et al. (2008).

In information terms one would expect that when a local less common configuration is encountered at a certain position i in the sequence of cells, the conditional probability for that configuration given the n-length symbol sequence, for example, in the cells to the left of it will be relatively small. This implies a high *local information* $I_{i,n}^{(L)}$ of that conditional probability,

$$I_{i,n}^{(L)} = \log \frac{1}{p(s_i | s_{i-n} \ldots s_{i-1})} , \tag{4.22}$$

where s_i denotes the symbol at position i, see Helvik et al. (2004, 2007). A corresponding local information $I_{i,n}^{(R)}$ conditioned on the n cells to the right of the position i is similarly defined. If all correlations are included, i.e., in the limit $n \to \infty$, such a local information quantity has a spatial average that can be written

$$\lim_{n\to\infty} \langle I_{i,n}^{(L)} \rangle = \lim_{N\to\infty, n\to\infty} \frac{1}{2N+1} \sum_{i=-N}^{N} \log \frac{1}{p(s_i | s_{i-n} \ldots s_{i-1})} . \tag{4.23}$$

By using the ergodicity theorem, Eq. (3.7), we find that

$$\begin{aligned} \lim_{n\to\infty} \langle I_{i,n}^{(L)} \rangle &= \lim_{n\to\infty} \sum_{x_0 \ldots x_n} p(x_0 \ldots x_n) \log \frac{1}{p(x_n | x_0 \ldots x_{n-1})} = \\ &= \lim_{n\to\infty} \Delta S_{n+1} = s . \end{aligned} \tag{4.24}$$

The same holds for the corresponding right-sided quantity $I_{i,n}^{(R)}$. We can define a local information as an average between the two,

$$I_{i,n} = \frac{1}{2} \left(I_{i,n}^{(L)} + I_{i,n}^{(R)} \right) , \tag{4.25}$$

This means that the local information $I_{i,n}$, as well as the left- and right-handed versions, in the limit $n \to \infty$, has a spatial average that equals the entropy of the system.

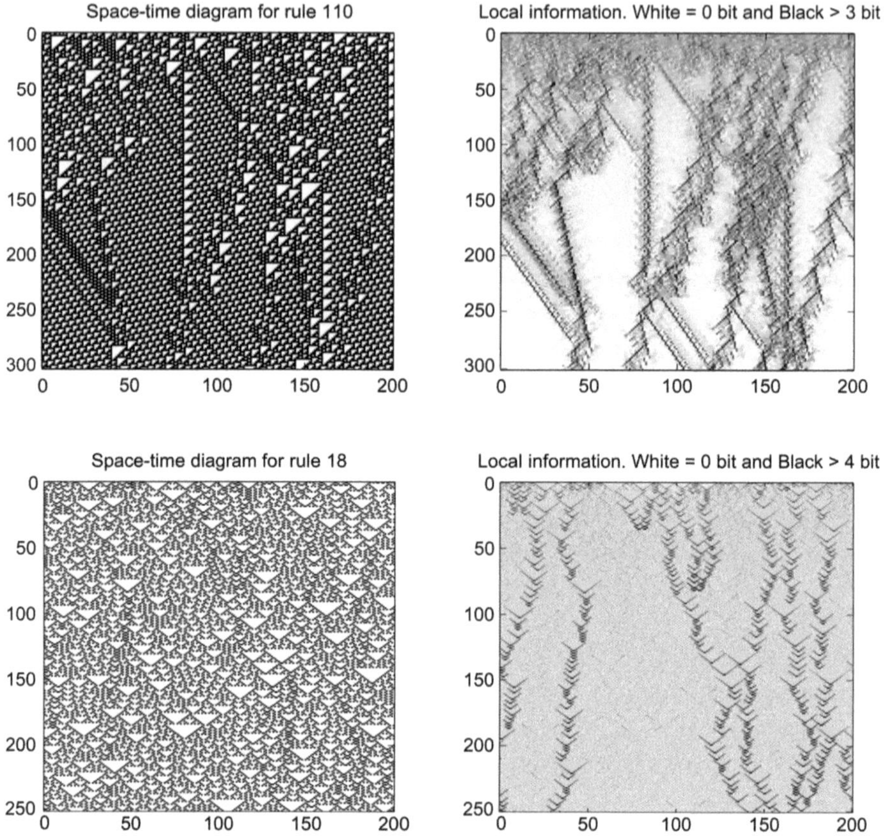

Fig. 4.14 Top row: The space-time CA pattern for rule R110 (left), and the corresponding picture of the local information $I_{i,n}$ (right). Here, $n = 14$ and the system length is 10^7. (The grey scale from light to dark spans from low to high information values.) The local information picture clearly shows the presence of spatial configurations that deviate from the dominating background pattern. Bottom row: Corresponding pictures for the rule R18. Here, the local information picture reveals that there are rare deviations from an irregular background. In this case there are pairs of 1's that become less and less common in the time evolution and therefore contribute to a higher local information value. Figure from Helvik et al. (2004)

$$\lim_{n \to \infty} \langle I_{i,n} \rangle = s \ . \tag{4.26}$$

In Fig. 4.14, the local information is applied to the patterns generated by two cellular automaton rules: the class IV rule R110 and the irreversible class III rule R18. Rule R110 creates a background pattern with a periodic structure in both space and time, and the local information in this part is decreasing over time, while the propagating structures do carry some information as is clearly seen in the figure. For rule R18, the figure exemplifies that there may be local information that is high, reflecting less common local configurations, not easily detected in the chaotic pattern of R18. The

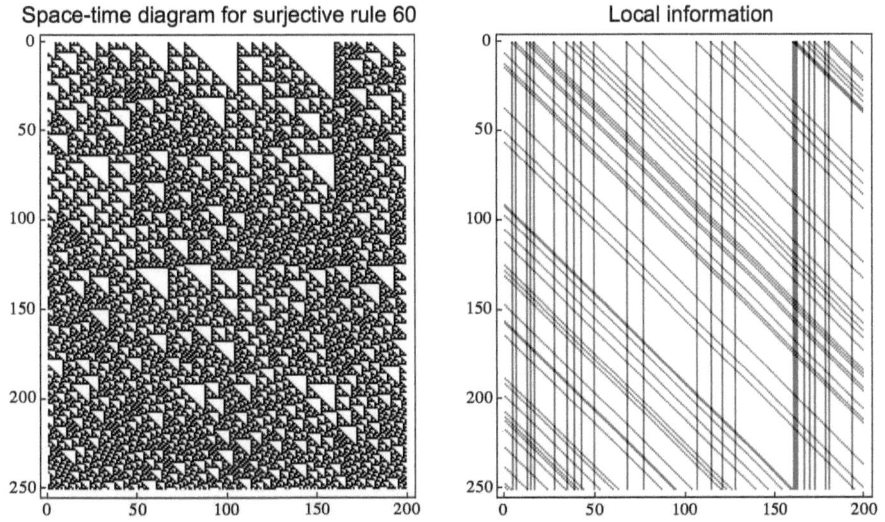

Fig. 4.15 The space-time CA pattern for rule R60, and the corresponding local information picture. Here the local information $I_{i,\infty}$ is derived analytically. A good numerical estimate would not be computationally possible because of the constantly increasing correlation lengths. (Grey scale coding as in Fig. 4.14.) Figure from Lindgren (2015)

information in this case stems from the presence of pairs of 1's that become more and more rare in the time evolution.

In Fig. 4.15 the space-time pattern and the local information is shown for the almost reversible class III rule R60. In this case, the local information is analytically derived. The specific structure of this almost reversible rule implies, when the state 0 dominates initially, that cells with the state 1 give rise to a high local information, both at the initial position where it remains and at a position that propagates to the right over time. This is connected to the discussion we had on rule R60 regarding its reversibility property and the constantly increasing correlation length.

Problems

4.1

Suppose that the initial state of a cellular automaton is generated by the following automaton, where the probability for choice of arc is $1/2$.

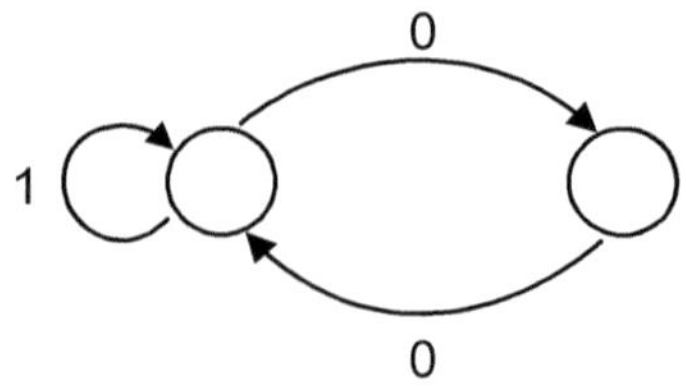

How large is the entropy initially? If the rule is R18, how does the automaton look like that describes the system after one time step, and what is the entropy s? What will the entropy be at this time step if the rule instead is R22?

4.2 CA entropy

Consider a one-dimensional cellular automaton given by elementary rule R71 (where configurations 110, 010, 001, and 000 result in a 1 and the others give 0). Let the initial state be characterised by the following finite state automaton

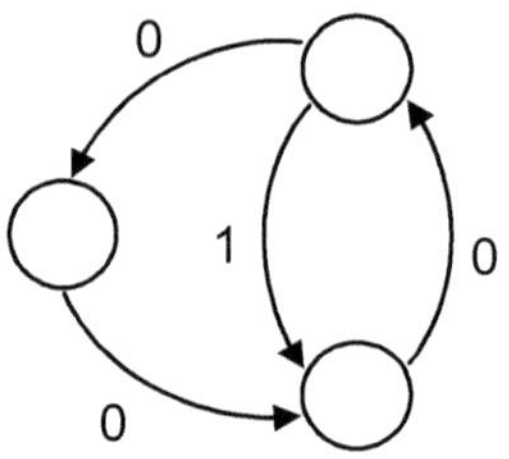

where the probabilities for choosing an arc is always the same (1/2) if there is a choice. What is the initial entropy ($t = 0$), and what is the entropy at $t = 1$ and $t = 2$?

4.3 Lattice gas entropy

Consider an infinite 2-dimensional lattice gas constructed in the following way. The space is a square lattice and in each cell up to four particles may be present (one in each direction). The system evolves in discrete time, and in each time step there is movement and collision. Particles move from one cell to the next according to the direction of the particle. A collision occurs if and only if exactly two particles enter a cell with opposite directions, and then the direction of these particles are changed so that they leave perpendicular to their initial directions. The two processes in a single time step is illustrated in the following figure.

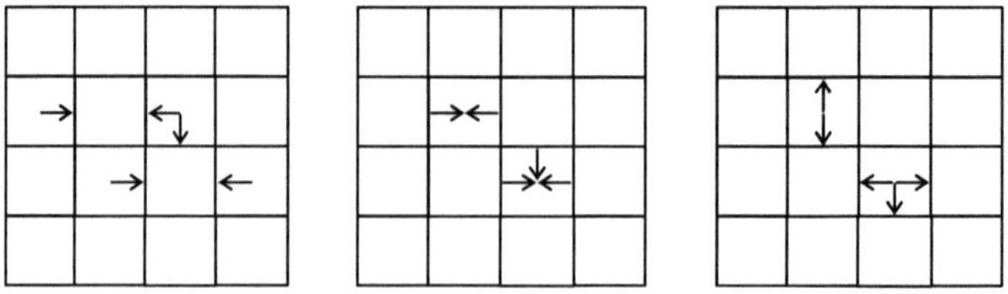

Suppose we have a system where we initially (at $t = 0$) have equal densities of the four particle directions ($\rho/4$ each, with ρ being the overall particle density), but where particles initially are present only in cells where all particle directions are present. Assume that these cells, each containing four particles, are randomly distributed over the whole lattice.

Consider the entropy s of the spatial configuration of particles, based on the 2-dimensional block entropy, $s = S_{m\times m}/m^2$ in the limit infinitely large blocks, $m \to \infty$, see Eq. (3.44).

What is the entropy s at $t = 0$? How does the entropy change in the time evolution? If one would estimate the entropy using a finite block size m after very long time T, with $T >> m$, what result should one expect? What is the explanation?

4.4 CA entropy

Consider a one-dimensional cellular automaton given by elementary rule R128 (where configuration 111 results in a 1 and the others result in 0). Let the initial state be characterised by the following finite state automaton

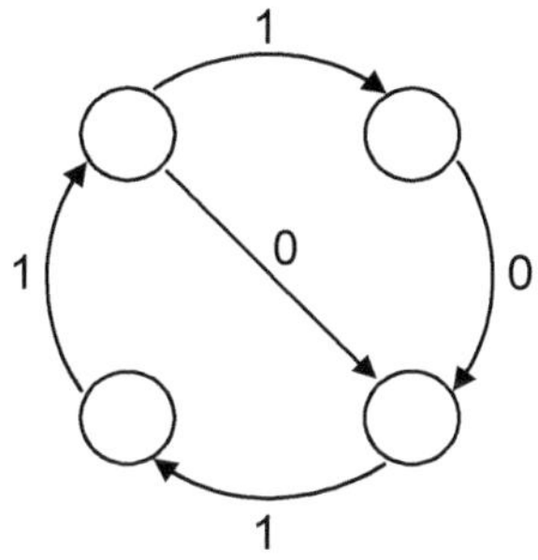

where the probabilities for choosing an arc is always the same (1/2) if there is a choice. Determine the finite state automaton that characterises the state at time $t = 1$. What is the initial entropy ($t = 0$), and what is the entropy at $t = 1$ and $t = 2$?

4.5 CA entropy

Consider a one-dimensional cellular automaton given by elementary rule R192 (where configurations 111 and 110 result in a 1 and the rest give 0). Let the initial state be characterised by the following finite state automaton

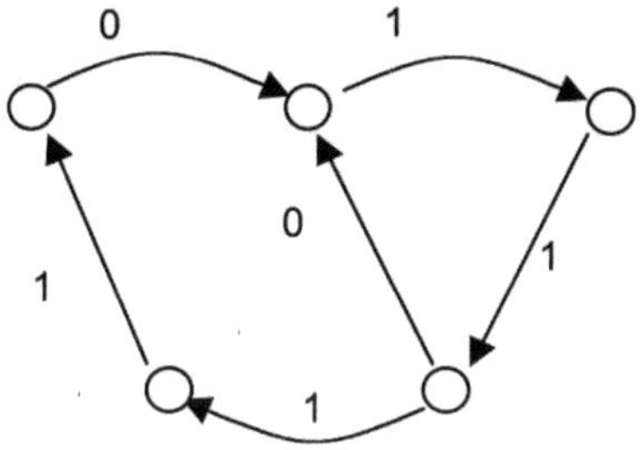

where the probabilities for choosing an arc is always the same (1/2) if there is a choice. Determine the finite state automata that characterise the state at time $t = 1$ and at time $t = 2$, respectively. What is the initial entropy ($t = 0$), and what is the entropy at $t = 1$, $t = 2$, and $t = 3$?

4.6
Suppose that the initial state ($t = 0$) for the elementary CA rule R18 is described by the finite automaton with transition probabilities q and $1 - q$:

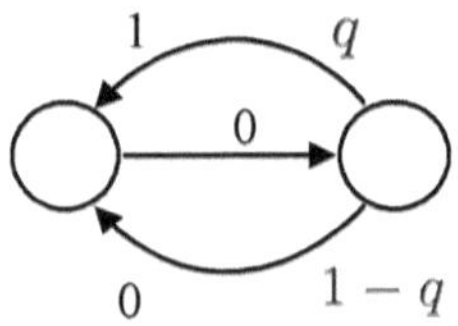

First, suppose that $q = 1/2$. What does the finite automaton look like that describes the state at time $t = 1$? How has the entropy changed between these time steps? What is the entropy at $t = 2$?

If $q > 1/2$, what is then the entropy s at $t = 2$?

4.7
Consider a binary periodic sequence ...01010101... . If noise is added so that symbols in this sequence are flipped ($0 \to 1$ and $1 \to 0$) with a certain probability $q < 1/2$, how does the correlation complexity η change?

4.8
Suppose that elementary rule R129 governs the time evolution of a cellular automaton, and that the initial state (at $t = 0$) is characterised by the following finite automaton

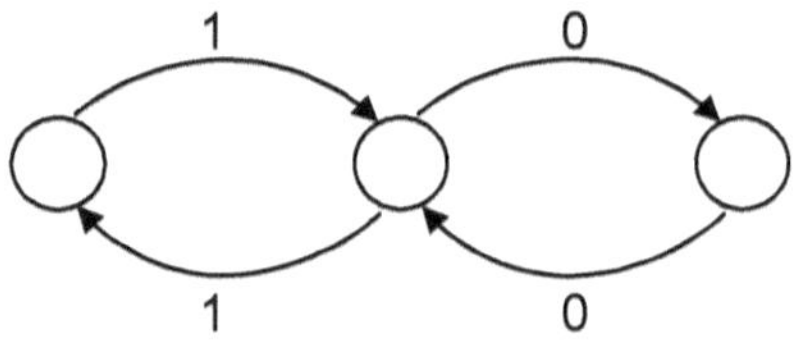

where the probabilities for choosing 0 and 1 from the middle node are equal (1/2). What is the entropy s after one time step ($t = 1$) and after two steps ($t = 2$)? Suppose instead that the probabilities for choosing 0 and 1 in the middle node differ, say $p(0) < 1/2 < p(1)$. Describe in a qualitative way how this affects entropy and correlations at the first time step ($t = 1$).

4.9
Suppose that elementary rule R68 governs the time evolution of a cellular automaton, and that the initial state (at $t = 0$) is characterised by the following finite automaton

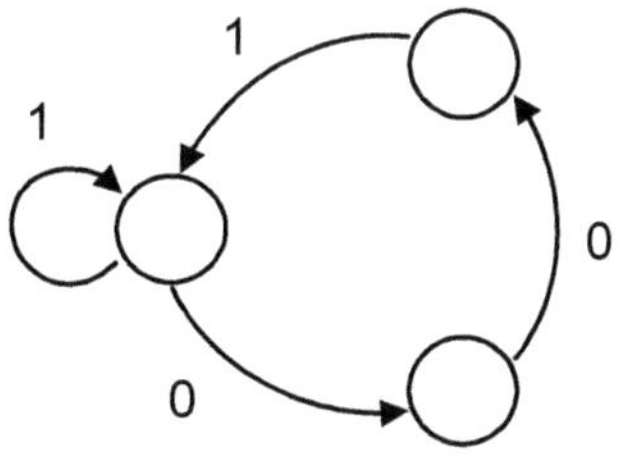

where the probabilities for choosing 0 and 1 from the bottom left node are equal (1/2). What is the initial entropy, and how does it change over time?

Chapter 5
Physics and Information Theory

Abstract This chapter starts with a short introduction of some basic thermodynamic concepts and laws. This is followed by an example involving the mixing of ideal gases, where the change in information is examined alongside thermodynamic properties such as entropy change and work. This serves to highlight the link between entropy in information theory and thermodynamics. In statistical mechanics, a physical system can be characterised by certain macroscopic properties, expressed as expectation values for variables like the internal energy of the system or the number of molecules of a certain type. We demonstrate how the maximum entropy formalism can be applied to such a situation to establish the probability distribution that characterises the system—the Gibbs distribution. It is shown how the formalism can be used to derive a number of thermodynamic relations. As a specific example, for which there is a strong connection to information theory, we explore one-dimensional spin systems. Such a system can be regarded as a symbol sequence generated by a stochastic process, allowing us to directly employ the methodology from Chap. 3. The entropy based on the internal statistics from such a symbol sequence corresponds to the statistical mechanics entropy of the system (factoring in Boltzmann's constant). Consequently, we can utilise the maximum entropy formalism on the symbol sequence in order to determine the equilibrium characteristics of the corresponding spin system. The one-dimensional Ising model serves as an illustrative example. Finally, we make an information characterisation of a reversible spin system dynamics in the form of a cellular automaton. It is demonstrated how the approach towards equilibrium can be understood despite the fact that microscopic reversibility conserves the entropy of the system.

5.1 Basic Thermodynamics

When observing, or measuring on, a physical system we often gain knowledge on some macroscopic characteristics of the system, for example its energy or molecular composition. We seldom have full information on the exact microscopic configuration of the system, but we may, based on what we know regarding the macroscopic

© Springer-Verlag GmbH Germany 2024

K. Lindgren, *Information Theory for Complex Systems*, Understanding Complex Systems, https://doi.org/10.1007/978-3-662-68214-2_5

properties, assign a probabilistic description of the system. Such a description could be in the form of a probability distribution over the possible microstates of the system. Following the discussion in Chap. 2, such an assignment of probabilities should not include more knowledge than we have, and thus we should use a maximum entropy approach.

Before describing in detail how the maximum entropy approach connects information theory with statistical mechanics, we will study a simpler example in which the information is directly connected to the work that can be extracted out of an ordered, non-equilibrium, system, when bringing that system reversibly to an equilibrium, i.e., to a fully disordered state. That example will be based on some basic thermodynamic definitions and relations that will first be summarised below.

In the following section we will consider ideal gases (pure or mixed gases). In an ideal gas, molecules are assumed to be point particles that do not interact. For such a system the ideal gas law holds,

$$pV = Nk_{\mathrm{B}}T \,. \tag{5.1}$$

Here, p is the pressure, V the volume, N the number of molecules in the gas, T the temperature, and k_{B} Boltzmann's constant (1.38×10^{-23} J/K).

We will use this relation in order to determine the work that is required (or that can be extracted) when a gas volume is changed. If one assumes that the system may interact with the environment (the world outside the system defined by the volume V) by receiving energy in the form of *heat* Q or deliver energy in the form of *work* W, then we can use Fig. 5.1 to illustrate the first law of thermodynamics.

The change ΔU of *internal energy* U must equal the difference between heat received and work delivered. (Heat and work may, of course, go in the other direction, when the corresponding variable is negative.) This is *the first law of thermodynamics* stating that *energy is a conserved quantity*,

$$\Delta U = Q - W \,. \tag{5.2}$$

If the work is pressure-volume work, i.e., work $\mathrm{d}W$ extracted from an ideal gas from an infinitesimal volume change $\mathrm{d}V$, then the work is

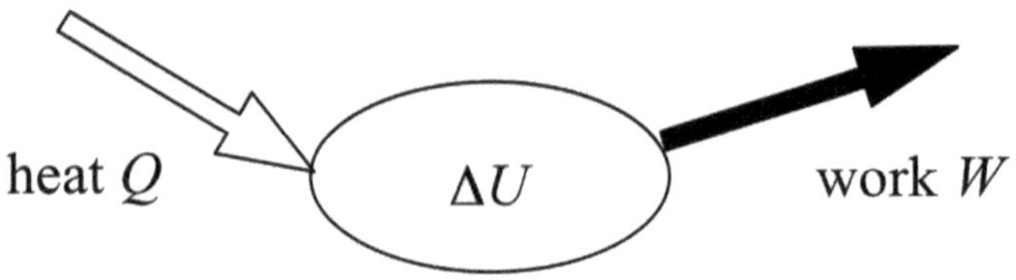

Fig. 5.1 The first law of thermodynamics states that energy is a conserved quantity. The change in internal energy U is determined by the net energy difference resulting from the inflow of heat Q and work extracted from the system W

$$\mathrm{d}W = p\,\mathrm{d}V\,. \tag{5.3}$$

Thermodynamic entropy S_T is a quantity that follows a heat flow but that is not present in the energy flow associated with the work. We will use the notation S_T for thermodynamic entropy to distinguish that from the information-theoretic entropy. As will be seen, there is a simple relation between the two. With a heat flow $\mathrm{d}Q$ of temperature T follows an entropy increase of the system $\mathrm{d}S_\mathrm{T}$,

$$\mathrm{d}S_\mathrm{T} = \frac{\mathrm{d}Q}{T}\,. \tag{5.4}$$

This is assuming processes in which the change of the system state is arbitrarily close to a local equilibrium. In general, the entropy change is at least as large as is stated in Eq. (5.4), where an additional increase may come from irreversible processes driving the system towards equilibrium.

Similarly, a heat flow leaving a system leads to a decrease in the entropy of the system (but increases the entropy in the system that receives the heat). For a closed system, the *second law of thermodynamics* states that *the total entropy* $S_\mathrm{T,total}$ *always increases or stays constant*,

$$\mathrm{d}S_\mathrm{T,total} \geq 0\,. \tag{5.5}$$

For, example, if one part of a system has higher temperature T_H and loses heat $\mathrm{d}Q$ to a colder part with temperature $T_\mathrm{C} < T_\mathrm{H}$, there is an entropy decrease of the first part, $dS_\mathrm{H} = \mathrm{d}Q/T_\mathrm{H}$ which is then smaller than the corresponding increase of the second part, $dS_\mathrm{C} = \mathrm{d}Q/T_\mathrm{C}$, and the total entropy change is then positive, $\mathrm{d}S_\mathrm{T,total} = -dS_\mathrm{H} + dS_\mathrm{C} = \mathrm{d}Q(1/T_\mathrm{C} - 1/T_\mathrm{H}) > 0$.

5.1.1 Intensive and Extensive Variables

We will consider a system of volume V that contains M different molecular species, each with a certain number of molecules N_i ($i = 1, \ldots, M$). Further, we assume that the system is characterised by a certain internal energy U. If we assume that the system is in internal equilibrium, the thermodynamic entropy S_T of the system is well defined by the state variables U, V, and N_i.

These state variables are extensive variables, i.e., they increase linearly with system size. One can characterise the same system using intensive variables instead, related to these extensive variables. Sometimes the relations below, based on the how the thermodynamic entropy S_T depends on the extensive variables, are used as definitions of the intensive variables *temperature* T, *pressure* p, and *chemical potential* g_i. The relations are partial derivatives of the thermodynamic entropy S_T with respect to one of the extensive variables (U, V, or N_i), with the other variables kept constant (as indicated by the notation),

$$\left(\frac{\partial S_{\mathrm{T}}}{\partial U}\right)_{V,N_i} = \frac{1}{T}\,, \tag{5.6}$$

$$\left(\frac{\partial S_{\mathrm{T}}}{\partial V}\right)_{U,N_i} = \frac{p}{T}\,, \tag{5.7}$$

$$\left(\frac{\partial S_{\mathrm{T}}}{\partial N_j}\right)_{U,V,N_{i\neq j}} = -\frac{g_j}{T}\,. \tag{5.8}$$

We will use these relations when interpreting the results from using the maximum entropy approach from information theory within statistical mechanics in Sect. 5.3.

5.2 Work and Information—An Extended Example

In this section we consider the following experiment dealing with mixing of ideal gases. First, consider two gases that are separated within a container by a wall. When the wall is removed the gases expand into the full volume of the container, and finally the system consists of the mixture of the two gases occupying the full volume. In the mixed system we have less information about the positions of gas molecules compared to the case when the gases are separated by the wall. It turns out that the additional information we have in the separated case is directly related to the amount of work that we can extract when bringing the system to the mixed state in a controlled way. Below we shall see how the information is related to the energy that can be extracted in the form of work. This will also give us a relation between the information-theoretic entropy describing the disorder of the mixed system and the corresponding thermodynamic entropy.

Let us consider a gas container of volume V with a mixture of two different ideal gases 1 and 2, see Fig. 5.2. The total number of molecules is N, of which N_1 are of type 1 and N_2 are of type 2. The normalised concentrations can be written $x_1 = N_1/N$ and $x_2 = N_2/N$. If we pick a molecule from a certain part of the container, we do not know which molecule we will get, but we describe the chances by the probability distribution given by the concentrations, $P = \{x_1, x_2\}$. The information-theoretic entropy S can then be used to quantify our lack of knowledge about the system (which molecule we will get),

$$S = x_1 \log \frac{1}{x_1} + x_2 \log \frac{1}{x_2}\,. \tag{5.9}$$

Consider now the case where the two gases are separated. The container is divided into two parts with volumes V_1 and V_2, so that the pressure is the same everywhere, implying $V_1 = x_1 V$ and $V_2 = x_2 V$, see Fig. 5.2. If we now pick a molecule from a certain place, we know which molecule it is and our uncertainty (entropy) is zero. By mixing the gases, we make an information loss of $\Delta S = S$ per molecule.

Fig. 5.2 The two gases are mixed in the container to the left, but to the right they are sorted and separated by a wall so that the pressure is equal in the two parts

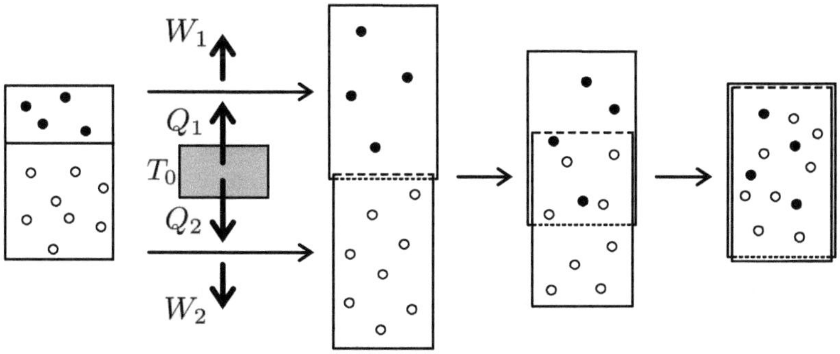

Fig. 5.3 The two gases are mixed reversibly in two steps: First, an isothermal expansion is done by taking up heat, $Q = Q_1 + Q_2$ from a heat reservoir with temperature T_0, at the same time as work is done by the expansion, $W = W_1 + W_2$. This is followed a fully reversible mixing, using semi-permeable membranes, without any energy transfer

We will now calculate the maximum amount of work that can be derived from a process that mixes the two gases. Then we may compare that with the information loss we make by mixing the gases. We will also derive the thermodynamic increase of entropy, to be compared with the information-theoretic one.

Suppose that the gas container is in thermodynamic equilibrium with the environment at temperature T_0. Since the initial volume and the final volume of our system are equal, there will be no net pressure-volume work done on the environment and we may assume that the environmental pressure is zero in our calculations. Now we shall mix the gases in a reversible way as follows. First, let the separated gas volumes expand isothermally, each to the full volume V, by allowing a heat flow from the environment to keep the temperature constant, see Fig. 5.3. Next, we assume that the two volumes can slide into each other as indicated in the figure, and that the two walls that keep the molecules within the their respective volume V are semi-permeable, so that only the other molecular type can freely pass. This allows us to mix the gases reversibly by letting them pass the semi-permeable walls, when the two volumes are pushed together. Note that there is no net force involved in this mixing, since the molecules in the first volume do not recognise the wall of the second, and vice versa. There is no heat flow involved in the mixing, so the entropy is unchanged in this part of the process. The only work (and entropy change) in this process comes from the expansion *before* the mixing takes place. By using the ideal gas law, the work W_i, for $i = 1$ and 2, is determined by Eqs. (5.1) and (5.3),

$$W_i = \int_{V_i}^{V} p_i \mathrm{d}v = \int_{V_i}^{V} N_i k_\mathrm{B} T_0 \frac{\mathrm{d}v}{v} = N_i k_\mathrm{B} T_0 \ln \frac{V}{V_i} = N k_\mathrm{B} T_0 x_i \ln \frac{1}{x_i} , \qquad (5.10)$$

which results in

$$W = W_1 + W_2 = N k_\mathrm{B} T_0 \left(x_1 \ln \frac{1}{x_1} + x_2 \ln \frac{1}{x_2} \right) = N k_\mathrm{B} T_0 (\ln 2) S , \qquad (5.11)$$

where the factor (ln 2) enters since the information-theoretic entropy S is defined by log of base 2, as in Eq. (5.9). The relation between extractable work per molecule w and information loss when the system is mixed ΔS is then

$$w = k_\mathrm{B} T_0 (\ln 2) \Delta S . \qquad (5.12)$$

Next, we would like to relate the information-theoretic increase in disorder, ΔS, with the thermodynamic increase in entropy ΔS_T. We have assumed an isothermal process (i.e., constant temperature). Since the internal energy U of an ideal gas is a function of temperature only (since there is no interaction between the molecules), energy conservation implies that the work W done by the system must equal the heat flow Q into the system. The change of thermodynamic entropy $\Delta S_T = Q/T_0$ for the whole system can then be written

$$\Delta S_T = \frac{Q}{T_0} = N k_\mathrm{B} (\ln 2) \Delta S . \qquad (5.13)$$

This illustrates the fact that the conversion factor between information-theoretic S and thermodynamic entropy per molecule s_T is $k_\mathrm{B}(\ln 2)$,

$$s_\mathrm{T} = k_\mathrm{B} (\ln 2) S . \qquad (5.14)$$

Note that the information-theoretic quantity here is also expressed on the level of a single molecule, since it quantifies information associated with observing one molecule from the system. One bit of information thus has a very low thermodynamic value. This example also illustrates that the information loss when mixing two gases is not really a result from a mixing process, but it comes from the fact that each gas, after the mixing, is distributed over a larger volume.

5.3 From Information Theory to Statistical Mechanics and Thermodynamics

Information theory can be used as a tool to determine the probability distribution over microstates that is consistent with the known macroscopic properties of a physical system, like energy and number of particles. Sometimes these macroscopic variables are said to define the macrostate of the physical system. Here we aim for a description of the physical system in internal equilibrium given the known macroscopic variables, and we will use the word *macrostate* for the probability distribution over microstates.

According to the maximum entropy principle, we should choose the macrostate that has the largest entropy and that is consistent with the known variables. The application of the maximum entropy principle to statistical mechanics and thermodynamics was developed by Jaynes (1957). For a comprehensive presentation of physics and information, with several good examples and illustrations, see Bais and Farmer (2007). In this section, which builds on the work in Eriksson et al. (1987), we demonstrate the connection between concepts in information theory and concepts in statistical mechanics and thermodynamics.

We consider a system characterised by a certain volume V, and M different molecular species. The system is part of a larger surrounding equilibrium system with a possibly different energy per unit volume as well as different concentrations n_i $(i = 1, \ldots, M)$, as indicated in Fig. 5.4. Our knowledge about the studied system of volume V is that its composition is given by expectation values (or averages) N_i $(i = 1, \ldots, M)$ of the number of molecules of the different species. We also know the the expectation value U of the internal energy.

This is the type of situation discussed in the Sect. 2.2 on the maximum entropy formalism. We do have some knowledge about certain properties of the system in terms of averages, or expectation values, but we do not know the microstate, nor the probability distribution over the microstates. The maximum entropy formalism can now be applied to derive a distribution over the microstates. Like in Sect. 2.2, the maximisation problem can be formulated as

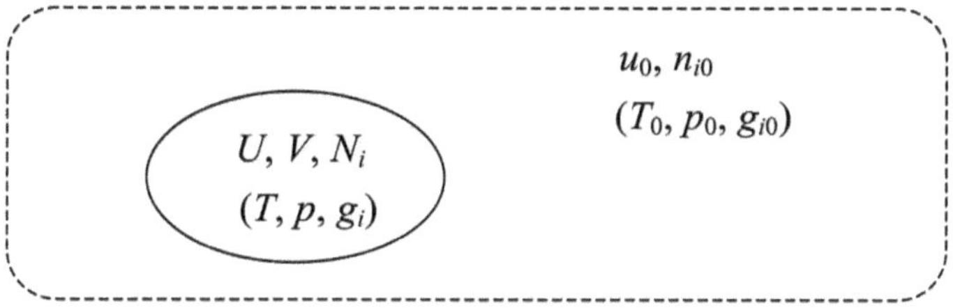

Fig. 5.4 We consider a system of volume V, with an expected internal energy, U, and M different molecular species, each with an expected number of molecules N_i. The system is placed in a large environment characterised by a certain energy density u_0 and certain molecular densities (concentrations). Both the system and the environment can also be characterised by their corresponding intensive variables, temperature T, the pressure p, and the chemical potentials g_i $(i = 1, \ldots, M)$

Choose $P = \{p_i\}_{i \in \text{microstates}}$ so that the entropy

$$S[P] = \sum_i p_i \ln \frac{1}{p_i} \tag{5.15}$$

is maximised, subject to constraints on, (i) energy,

$$\sum_i p_i h(i) = U \ , \tag{5.16}$$

(ii) number of molecules of each species,

$$\sum_i p_i f_k(i) = n_k \ , (k = 1, \ldots, M) \ , \ \text{and} \tag{5.17}$$

(iii) normalisation,

$$\sum_i p_i = 1 \ . \tag{5.18}$$

Here the microscopic function $h(i)$ determines the energy for each microstate i, and in average that should be equal to the internal energy U. Similarly, the functions $f_k(i)$ determine the number of molecules of type k in microstate i, which in average should be equal to N_k. In order to solve the general problem and to derive some important relations, this level of description is sufficient, and we do not need to specify these functions further.

Following the Lagrangian formalism in Sect. 2.2, we introduce Lagrangian variables: β for the energy constraint, $\lambda_k (k = 1, \ldots, M)$ for the M different constraints on number of molecules, and $\mu - 1$ for the normalisation constraint. The solution is then a Gibbs distribution, cf., Eq. (2.22),

$$p_i = \exp\left(-\mu - \beta h(i) - \sum_{k=1}^{M} \lambda_k f_k(i)\right) \ , \tag{5.19}$$

The entropy of the Gibbs distribution, expressed in the Lagrangian variables, is easily determined,

$$S = \mu + \beta U + \sum_{k=1}^{M} \lambda_k N_k \ . \tag{5.20}$$

At this point, we can determine how the Langrangian variable μ depends on volume V. If we scale up the system by a certain factor, increasing all extensive variables (energy, number of molecules), for example by putting together two identical systems (i.e., multiplying with a factor of 2), then all variables S, U, and N_k, will increase with the same factor. Behind the linear increase also in entropy is an assumption that the microstates of the systems that are combined are uncorrelated (or that the systems are independent), which means that we consider systems that are very large compared to the interaction distance between molecules that, for example, determine the energy of a microstate. Then we know from the additivity of the entropy, see Eq. (2.11), that the entropy scales linearly with the size.

This means that also μ must have such a linear dependence on volume: $\mu = \mu(V) = \mu_1 V$, where μ_1 does not depend on volume. If we now also use the relation between thermodynamic entropy and the information-theoretic one (with information expressed in natural units using the natural logarithm), $S_T = k_B S$, we can write Eq. (5.20) as,

$$\frac{S_T}{k_B} = \mu_1 V + \beta U + \sum_{k=1}^{M} \lambda_k N_k \,. \tag{5.21}$$

By using this equation together with the thermodynamic definitions on temperature T, pressure p, and chemical potential g_i, from Eqs. (5.6)–(5.8), we find that the Lagrangian variables can be expressed in these intensive variables,

$$\left(\frac{\partial S_T}{\partial U}\right)_{V,N_i} = k_B \beta = \frac{1}{T} \quad \Rightarrow \quad \beta = \frac{1}{k_B T} \,, \tag{5.22}$$

$$\left(\frac{\partial S_T}{\partial V}\right)_{U,N_i} = k_B \mu_1 = \frac{p}{T} \quad \Rightarrow \quad \mu_1 = \frac{p}{k_B T} \,, \tag{5.23}$$

$$\left(\frac{\partial S_T}{\partial N_j}\right)_{U,V,N_{i \neq j}} = k_B \lambda_j = -\frac{g_j}{T} \quad \Rightarrow \quad \lambda_j = -\frac{g_j}{k_B T} \,. \tag{5.24}$$

By inserting this in Eq. (5.21), we get the *Gibbs equation*,

$$U = T S_T - pV + \sum_{k=1}^{M} g_k N_k \,. \tag{5.25}$$

By using the maximum entropy formalism to derive a probability distribution over microstates, the Gibbs distribution, we have derived one of the fundamental relations in thermodynamics.

5.3.1 Comparing Two Different Gibbs Distributions

There are often reasons for comparing two different Gibbs distributions, for example, when a smaller physical system, characterised by a certain Gibbs distribution P, deviates from an environment system characterised by the distribution P_0 that would describe the system if it would be in equilibrium with the environment. In that equilibrium situation, the system would have other values on its extensive variables (energy and number of molecules of different kinds). With the notation from Fig. 5.4, we would have for the internal energy $U_0 = u_0 V$, for the number of molecules $N_k = n_k V$. The environment (and the system in equilibrium with it) would also be characterised by intensive variables, the temperature T_0, the pressure p_0, and the chemical potentials g_{k0} $(k = 1, \ldots, M)$. Again, using the maximum entropy formalism to derive the distribution describing the equilibrium situation, we get the distribution P_0, characterised by the probabilities

$$p_{i0} = \exp\left(-\mu_0 V - \beta_0 h(i) - \sum_{k=1}^{M} \lambda_{k0} f_k(i)\right) . \tag{5.26}$$

Since the constraints for the equilibrium distribution is different, we have other Lagrangian variables now, as indicated by the subscript 0. Like in Eqs. (5.22)–(5.24), we can replace the Lagrangian variables for the equilibrium situation with the corresponding intensive thermodynamic variables,

$$\beta_0 = \frac{1}{k_B T_0} , \tag{5.27}$$

$$\mu_0 = \frac{p_0}{k_B T_0} , \tag{5.28}$$

$$\lambda_{j0} = -\frac{g_{j0}}{k_B T_0} . \tag{5.29}$$

One can now ask the question: If we a priori assume that the system (in volume V) will be in equilibrium, characterised by the Gibbs distribution of Eq. (5.26), but after observation learn that the correct description is another Gibbs distribution, Eq. (5.19), how much information have we gained? This represents a situation where the observed system is not in equilibrium with the environmental system. Below we will relate such an information gain with thermodynamic properties of the non-equilibrium situation.

The relative information $K[P_0; P]$ between the a priori equilibrium description and the observed one, both being Gibbs distributions, can be written

$$\begin{aligned}
K[P_0; P] &= \sum_i p_i \ln \frac{p_i}{p_{i0}} = \\
&= \sum_i p_i \left((-\mu_1 + \mu_0)V + (-\beta + \beta_0)h(i) + \sum_{k=1}^{M}(-\lambda_k + \lambda_{k0}) f_k(i)\right) = \\
&= (\mu_0 - \mu_1)V + (\beta_0 - \beta)U + \sum_{k=1}^{M}(\lambda_{k0} - \lambda_k)N_k = \qquad (5.30) \\
&= \frac{1}{k_B T_0}\left(p_0 V + U - \sum_{k=1}^{M} g_{k0} N_k\right) - \frac{S_T}{k_B} , \qquad (5.31)
\end{aligned}$$

where we in the last step have used Eq. (5.21). If we multiply by $k_B T_0$, we get

$$k_B T_0 \, K[P_0; P] = U + p_0 V - T_0 S_T - \sum_{k=1}^{M} g_{i0} N_k \, . \tag{5.32}$$

It turns out that the expression on the right-hand side quantifies the amount of work that can be extracted from a process that brings the system characterised by the distribution P (or the extensive variables U, V, and N_i) into equilibrium with the environment characterised by P_0 (or the corresponding intensive variables T_0, p_0, and g_{i0}). Willard Gibbs introduced this general form of free energy already in 1875, see Dunbar et al. (1992). This free energy is often called the *exergy* of the system in the given environment. We will use the term exergy of a system as the most general form of free energy, which is then defined by the maximum amount of work that can be extracted when the system is reversibly brought to equilibrium with its environment. Thus, the exergy E equals the relative information between the equilibrium state and the actual one, multiplied with $k_B T_0$,

$$E = k_B T_0 \, K[P_0; P] \, . \tag{5.33}$$

By combining the expression for the exergy, Eq. (5.32), with the Gibbs equation (5.25) for internal energy, the exergy can be written

$$E = S_T(T - T_0) - V(p - p_0) + \sum_{i=1}^{M} N_i (g_i - g_{i0}) \, . \tag{5.34}$$

In this form it is clear that the exergy is zero when the system is characterised by the same intensive variables as the environment, which then implies that they are described by identical Gibbs distributions.

5.3.2 *Information and Free Energy in Non-equilibrium Concentrations*

Let us now introduce chemical concentrations, $c_i = N_i/V$ and $c_{i0} = N_{i0}/V$, for the system and its environment, respectively. Here we will focus on information and thermodynamic characteristics of a system that deviates from equilibrium in its concentrations of different molecules. Suppose, therefore, that the system has the same temperature and pressure as the environment, $T = T_0$ and $p = p_0$. Further, we assume that the chemical potential g_i, for chemical component i, can be written as for an ideal solution,

$$\frac{g_i}{k_B T} = C + \ln c_i \, . \tag{5.35}$$

Here, C is a constant. Then the exergy, Eq. (5.34), can be written as

$$E = \sum_{i=1}^{M} N_i(g_i - g_{i0}) = V k_B T_0 \sum_{i=1}^{M} \frac{N_i}{V} \ln \frac{c_i}{c_{i0}} =$$
$$= V k_B T_0 \sum_{i=1}^{M} c_i \ln \frac{c_i}{c_{i0}} . \qquad (5.36)$$

This is in a form very close to a relative information quantity, but the concentrations c_i and c_{i0} are not normalised. If we multiply them with a factor V/N the resulting quantities are normalised. Then we can rewrite the exergy in the form of a relative information between an a priori distribution over molecular types i given by $V c_{i0}/N$, and a new distribution $V c_i/N$,

$$E = k_B T_0 N \sum_{i=1}^{M} (c_i V/N) \ln \frac{c_i V/N}{c_{i0} V/N} = k_B T_0 N K[c_0 V/N; cV/N] , \qquad (5.37)$$

where N is the total number of molecules in the system, $N = \sum_k N_k$, and c and c_0 denote the concentration distributions.

The exergy is once again expressed in the form of a relative information between an a priori description and a new description. But in this case, the description is macrosopic, since we have a concentration distribution over molecular components, and not over microstates as was the starting point for the derivation of the exergy expression, cf. Eqs. (5.33) and (5.34). The normalised concentration distribution cV/N represents the probability $c_i V/N$ that a randomly selected molecule is of type i, and we can view this as a macroscopic representation of a chemical system. For example, if we make an image of a chemical system, the observed pattern can be assumed to be proportional to these concentration characteristics.

This relative information can serve as a starting point for examining spatial structure in chemical systems. It has the advantage that it is in an information-theoretic form at the same time as it connects to statistical mechanics and thermodynamics through the basic concept of exergy. The second law of thermodynamics tells us that, for a closed system, the entropy increases (or is constant), or, equivalently, that the exergy decreases (or stays constant). Such physical restrictions may be important in the description and analysis of chemical systems that exhibit self-organisation in the form of pattern formation. We shall return to such an analysis in a later chapter.

5.4 Microscopic and Macroscopic Entropy

The second law of thermodynamics originates from Clausius (1850) and Thomson (1852) (Lord Kelvin), who found a thermodynamic quantity which is non-decreasing in time. Clausius later introduced a term for this quantity: entropy (Clausius 1865). At this time, statistical mechanics had not been established, but there was a need for an understanding of the relation between thermodynamics and microscopic properties of gases. A first step was taken by Krönig (1856) who explored macroscopic phenomena in terms of microscopic properties. Soon Clausius, and later Maxwell and Boltzmann developed a kinetic theory for gases.

A problem which remained unsolved was how to understand the second law of thermodynamics within the framework of kinetic gas theory. In an effort to solve this, Boltzmann (1872) introduced the H-function, a microstate property of a gas, that under certain assumptions was proven to be decreasing in time until the system reaches an equilibrium given by the Maxwell–Boltzmann distribution law for molecular velocities. This is called the H-theorem, and the idea was that this should correspond to the second law of thermodynamics, with the H-function essentially being negative entropy. This approach met with difficulties, especially when it was applied to other systems than dilute gases, which led to the development of ensemble theory by Gibbs (1902) and Einstein (1902, 1903).

In ensemble theory, a large collection of systems is considered, each possessing certain known macroscopic properties. The microcanonical ensemble consists of systems with identical energy, whereas the canonical ensemble only defines an average energy for the entire ensemble. Each microstate in the ensemble is associated with a probability p_i, defining a distribution $P = \{p_i\}$ over the ensemble. Consequently, entropy is defined as $-k_{\mathrm{B}} \sum_i p_i \ln p_i$, i.e., an ensemble average of $-k_{\mathrm{B}} \ln p_i$. Unlike energy or the number of particles, which have both microscopic and macroscopic interpretations, entropy initially appears to be solely a macroscopic property.

However, in our analysis of entropy in symbol sequences, which draws on internal statistics of a long sequence (or microstate), we have regarded entropy as a measure of internal randomness or disorder. It follows that a microscopic equivalent of the macroscopic entropy should exist, characterising the thermodynamic properties of the system given sufficient internal statistics. Under general conditions, it can be demonstrated that for a system where each microstate can be represented by a symbol sequence σ, the entropy per symbol $s(\sigma)$ serves as a microscopic entropy (Lindgren 1988). In the thermodynamic limit, the ensemble average of the entropy $s(\sigma)$ for an individual microstate σ equals the thermodynamic entropy s_{T} per symbol (times a constant),

$$s_{\mathrm{T}} = k_{\mathrm{B}} \sum_{\sigma} p(\sigma) s(\sigma) \ , \tag{5.38}$$

where the summation is over all microstates in the ensemble. In fact, for almost all microstates in the ensemble, the microscopic entropy equals the thermodynamic entropy,

$$s_{\mathrm{T}} = k_{\mathrm{B}} s(\sigma) \ , \quad \text{for almost all } \sigma \ . \tag{5.39}$$

This can be understood if one realises that each microstate contains all correlations that are necessary for determining the whole ensemble. This is the requirement needed for the ergodicity theorem to hold, Eq. (3.7). The internal correlations of almost any microstate give restrictions to its randomness, expressed by its (internal) entropy $s(\sigma)$, as well as to the uncertainty of the macrostate, expressed by the thermodynamic entropy S_{T}.

In conclusion, the thermodynamic entropy—usually considered a measure of disorder or uncertainty regarding the specific microstate occupied by a physical system—also carries significance at the microscopic level. The microscopic entropy $s(\sigma)$ quantifies the internal disorder in the microstate σ based on internal correlations. When the physical system can be represented as a symbol sequence (or a d-dimensional discrete pattern of symbols), the information-theoretic framework discussed in Chap. 3 can be applied to calculate the internal entropy s, which corresponds to the system's thermodynamic entropy (up to a factor of Boltzmann's constant). The following sections will utilise this concept to analyse spin systems that can be represented by such a symbol sequence.

5.5 Information Theory for Spin Systems

In this section, we shall illustrate how the microscopic entropy can be used to determine the equilibrium state in simple one-dimensional spin systems.

In Sect. 5.3, we used the maximum entropy formalism to determine the macrostate, the probability distribution that characterises a certain physical system. Here we shall apply the maximum entropy formalism on the microscopic entropy of a spin system, or a system that can be described as a symbol sequence. This should then result in a probabilistic description of blocks of symbols in the microstate. Again, the maximisation is done under certain constraints, usually in the form of an average energy in the system. There are also constraints regarding typical properties that hold for probabilities of symbol sequences, for example, regarding summation over last or first index, see Eqs. (3.5) and (3.6).

In a spin system, the local interactions between spins (symbols) give rise to certain energy contributions, depending on the specific local configuration. This is usually characterised by an energy function that associates a certain energy value $h(x_1, \ldots, x_n)$ for each sequence of spins $x_1, \ldots, x_n$ within the interaction distance n. This captures the interaction energy between x_1 and the spins to the right (and in case of an external field, such an energy contribution for x_1 is also included). A given value of the internal energy u (per spin) in the system, then results in the constraint

$$\sum_{x_1, \ldots, x_n} p(x_1, \ldots, x_n) h(x_1, \ldots, x_n) = u \ . \tag{5.40}$$

The maximisation problem then takes the form: Find a set of probability distributions P_m over m-length sequences so that the entropy, Eq. (3.14),

$$s = \lim_{m \to \infty} \Delta S_m \tag{5.41}$$

is maximised, under the energy constraint, Eq. (5.40).

This may be very complicated, since in principle we need to determine an infinite series of probability distributions, P_m (with $m = 1, 2, \ldots$). The problem can be simplified, though, if there are no restrictions on longer sequences than interaction distance n, as stated by Eq. (5.40). Then, the entropy s equals ΔS_n, which means that we only need to consider sequences of length up to n. In order to see this, let us suppose that we get convergence only at ΔS_m, with $m > n$. This means that there is correlation information over blocks of length m, i.e., $k_m > 0$. But this is an unnecessary correlation information that reduces the maximum entropy, since we may reduce k_m to zero by selecting the m-block probability as $p(x_1, \ldots, x_m) = p(x_1, \ldots, x_{m-1}) p(x_m | x_2, \ldots, x_{m-1})$, without affecting the energy constraint. This procedure may then be repeated down to length $n + 1$, and for each step when correlation information is removed the entropy increases. So when searching for the maximum entropy distribution, it is sufficient to examine

$$\max \ \Delta S_n \ , \tag{5.42}$$

under the constraints above, and where n is the longest interaction distance entering in the energy constraint.

5.5.1 *Example: The One-Dimensional Ising Model*

As a more specific example, consider the one-dimensional Ising model (without an external field). This model was introduced by Ernst Ising (1925), in his attempt to model ferromagnetism, the spontaneous magnetisation for certain metals when temperature falls below a critical value. The original model was formulated in one dimension, and the material was modelled as a sequence of atoms, each having a certain magnetic spin, for simplicity assumed to point either upwards or downwards.

Therefore, suppose that our system is composed by an infinite sequence of spins,

$$\ldots \uparrow\downarrow\downarrow\downarrow\uparrow\downarrow\downarrow\uparrow\uparrow\uparrow\downarrow\downarrow\uparrow\uparrow\uparrow\uparrow\downarrow\downarrow\downarrow\uparrow\downarrow\uparrow\uparrow\downarrow\downarrow\downarrow\downarrow\downarrow\uparrow\uparrow\uparrow\uparrow\downarrow\uparrow\uparrow\downarrow\downarrow\downarrow \ldots ,$$

and that the energy in the system is determined by nearest neighbour interactions only: Configurations $\downarrow\downarrow$ and $\uparrow\uparrow$ contribute with energy $-J$, while $\uparrow\downarrow$ and $\downarrow\uparrow$ contribute with $+J$. Since h only depends on pairs of spins, it is sufficient to find maximum of ΔS_2, in order to determine the probability distributions. Then there are three probabilities to consider: $p_0 = P(\downarrow\downarrow)$, $p_1 = P(\downarrow\uparrow) = P(\uparrow\downarrow)$, and $p_2 = P(\uparrow\uparrow)$, where we have used the symmetry $P(\downarrow\uparrow) = P(\uparrow\downarrow)$ that must hold in an infinite binary symbol sequence. Probabilities for single spins can then be written $P(\downarrow) = p_0 + p_1$ and $P(\uparrow) = p_1 + p_2$.

To solve the maximisation problem, we introduce the Lagrange function

$$L(p_0, p_1, p_2, \lambda, \mu) = \Delta S_2 + \beta(u - J(2p_1 - p_0 - p_2)) \\ + \mu(1 - p_0 - 2p_1 - p_2) \,. \tag{5.43}$$

Here, β is the Lagrange variable related to the energy constraint, while μ is related to the normalisation constraint. The block entropy difference can be written

$$\Delta S_2 = p_0 \ln \frac{p_0 + p_1}{p_0} + p_1 \ln \frac{p_0 + p_1}{p_1} + p_1 \ln \frac{p_1 + p_2}{p_1} + p_2 \ln \frac{p_1 + p_2}{p_2} \,. \tag{5.44}$$

The derivation of the following solution is left as an exercise

$$p_0 = p_2 = \frac{1}{2\left(1 + e^{-2\beta J}\right)} \,, \tag{5.45}$$

$$p_1 = \frac{1}{2\left(1 + e^{2\beta J}\right)} \,. \tag{5.46}$$

Here we have chosen to keep the Lagrange variable β, instead of the energy u. As in the general Gibbs distribution, see Eq. (5.22), it is related to temperature, $\beta = (k_B T)^{-1}$. It is clear that in the zero temperature limit $T \to 0$, the system is characterised by larger and larger blocks of parallel spins, $p_0 = p_2$, while in the limit of $T \to \infty$, all p are equal, and we get a completely disordered state.

It is then clear, that with this model, we do not find ferromagnetism: There is no transition at a critical temperature under which the equilibrium solution has dominance of either spin direction. It is only in two or higher dimensions that the Ising model exhibits such a phase transition. The equilibrium solution in higher dimensions is a much more difficult problem, and the mathematical solution was not derived before the work by Onsager (1944).

5.6 The Approach Towards Equilibrium in a Reversible One-Dimensional Ising Model

We have seen how the equilibrium characteristics of spin systems can be derived by an information-theoretic procedure that seeks a microstate description in terms of correlations that gives the maximum entropy given the energy constraint of the system. An interesting question is in what way a microscopic dynamics starting from a non-equilibrium state can bring the system towards equilibrium and thus an increasing entropy. Is it possible to find such an equilibration also with a reversible microscopic dynamics? In principle, we have reversibility on the micro level, but then that implies that the (microscopic) entropy s stays constant. Only if the dynamics is probabilistic, like in a Monte Carlo simulation, there is an inflow of randomness that may lead to an increase in entropy.

In this section we present an analysis of a reversible micro dynamics, in the form of a cellular automaton rule, that leads to a modification of the spin configurations seemingly bringing the system closer to equilibrium. The aim is to explain what happens and in what way equilibrium can be said to be approached, despite the fact that microscopic reversibility implies a constant entropy s. This analysis is a summarised version of the investigation published in Lindgren and Olbrich (2017), where readers can find the mathematical details and proofs.

5.6.1 Q2R—A Microscopically Reversible Ising Dynamics

We use the Q2R model (Vichniac 1984) in one dimension, assuming an infinite system. This means that we describe the spatial state (infinite sequence of spins) at a specific time as the outcome of a stationary stochastic process, in the same way as we did in Chap. 4 on cellular automata. The system consists of an infinite sequence of states, either spin up or spin down, denoted by $\uparrow$ and $\downarrow$. Along with the spin state, a state also carries the information of whether it is updated or remains stationary in the current time step.

The updating structure mandates that every second spin is updated at time t, and the remaining half of the lattice gets updated in the next time step, continuing in this pattern. The updating rule inverts a spin when the flip does not alter the energy, and switches the state from updating to quiescent and vice versa. We assume a typical nearest neighbour Ising interaction, where parallel spins ($\uparrow\uparrow$ or $\downarrow\downarrow$) carry an energy of -1 and anti-parallel spins ($\uparrow\downarrow$ or $\downarrow\uparrow$) have an energy of $+1$. Consequently, the Q2R model becomes a cellular automaton simulation of the Ising model with conserved energy. The rule is also fully reversible, as will be seen below.

The initial state is generated by a stochastic process with distinct frequencies for $\uparrow$ and $\downarrow$, without correlations. We aim to conduct a statistical analysis of the temporal changes in spatial configurations. Each time step is thus characterised by a certain stochastic process, with the Q2R rule transforming this process from one time step to the next, in the same way as was done for cellular automata in Chap. 4.

The reversibility of Q2R is easily seen from exchanging the quiescent and updating states—an operation that flips the direction of time. Given that the Q2R rule is reversible, the entropy per position $s(t)$ of the ensemble at a particular time step t—equivalently, the entropy rate of the stochastic process that generates the ensemble at time step t—remains a conserved quantity under the Q2R dynamics. This follows from our discussion on cellular automata in Chap. 4.[1]

[1] This conservation of entropy is apparent given the existence of a local rule that operates inversely in time. As both of these rules are deterministic and inherently implies a non-increasing entropy, we can conclude that the entropy per position for an infinite system is preserved under the Q2R rule.

We further assume that the stochastic process exhibits spatial ergodicity. This implies that for almost every microstate within the ensemble at any given point in time, we have the sufficient statistics to compute any information-theoretic properties dependent on finite-length subsystems of the microstate. In effect, this allows us to characterise a single microstate at any time t, and identify its internal entropy and correlation characteristics. These features will be consistent with those of the entire ensemble. This notion is conceptually compelling as it enables us to identify the entropy as a property of a single microstate.

5.6.2 *Analysis of the Time Evolution Starting from a Low Entropy State*

It is assumed that the initial state, i.e., spin configuration, is generated by a stochastic process without correlations and with probability $0 < p < 1/2$ for spin up, $\uparrow$. Alongside this, a second binary variable indicates every second lattice site as being in an updating state ($\underline{0}$ or $\underline{1}$), leaving the rest in quiescent states (0 or 1). Here the direction of the spin is denoted by 0 and 1 (either underlined or not), signifying spin down and up, respectively. Hence, the probability of spin up is $p(\uparrow) = p(1) + p(\underline{1})$, and a similar calculation applies for spin down. The space-time pattern of the Q2R rule is shown in Fig. 5.5.

The entropy per site s, of such an initial state is simply the entropy of the stochastic process,

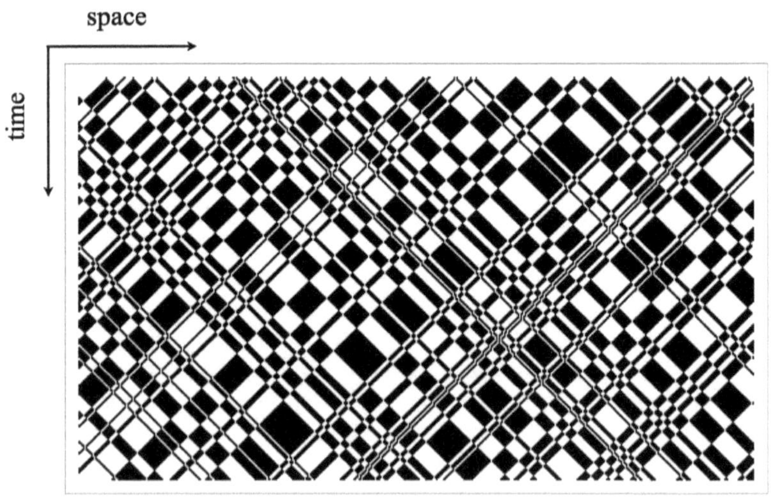

Fig. 5.5 The space-time pattern of Q2R with an initial state of high magnetisation (low frequency of the state $\uparrow$, or black cells)

$$s = S[\{p, 1-p\}] = p\log\frac{1}{p} + (1-p)\log\frac{1}{1-p}\;. \tag{5.47}$$

The alternating state structure of quiescent and updating states is perfectly ordered and therefore does not impact the entropy.

Taking into account the energy contributions of -1 and 1 from parallel and antiparallel spins, respectively, the energy density $u = -(1-2p)^2$ of the initial state is obtained. The system is not in equilibrium since the entropy s is not in a maximum given the energy density u. This is evident from the mere fact that the initial magnetisation is positive, as detailed in Sect. 5.5.1.

The core questions that this analysis and discussion aim to address are: Does the time evolution push the system closer to the maximum entropy description in any way? If so, how can it be explained in the context of conserved entropy s?

5.6.3 Time Evolution of the Magnetisation

The Q2R rule in one dimension translates to a straightforward addition modulo 2 rule for the updating states, $\underline{s}_{i,t}$, at position $i \in \mathbb{Z}$ and time t, while the quiescent states, $s_{j,t}$, simply copy the previous spin state. As a result, at time $t+1$, we get

$$\underline{s}_{i,t+1} = s_{i,t} \tag{5.48}$$

$$s_{j,t+1} = s_{j-1,t} + \underline{s}_{j,t} + s_{j+1,t} \;(\text{mod } 2)\;. \tag{5.49}$$

The addition modulo 2 for the updating states is an operation that flips the spin (0 or 1) whenever it does not affect the local energy. This operation allows us to express the local states at any time as a sum over initial spin states using the Proposition provided below (Lindgren and Olbrich 2017).

Proposition 5.1 *At time $t > 0$, updating and quiescent states, $\underline{s}_{i,t}$ and $s_{j,t}$, respectively, can be expressed as sums modulo 2 of initial spin states, $\xi_{i'}$ (with $i' \in \mathbb{Z}$), over certain intervals,*

$$\underline{s}_{i,t} = \sum_{i' \in I_{i,t-1}} \xi_{i'} \;(\textit{mod } 2)\;, \tag{5.50}$$

$$s_{j,t} = \sum_{j' \in I_{j,t}} \xi_{j'} \;(\textit{mod } 2)\;. \tag{5.51}$$

Here, $I_{i,n} = \{i-n, \ldots, i, \ldots i+n\}$ refers to the $(2n+1)$-length interval of positions centred around i. Consequently, a local updating state $\underline{s}_{i,t}$ at position i and time t depends on a sequence of $2t-1$ initial stochastic variables, while a quiescent state $s_{j,t}$ depends on a sequence of $2t+1$ initial stochastic variables.

The proof is contained in Appendix A.

This proposition infers that the distribution for local states, $P_1(t)$ or the single spin distribution, for $t > 0$ is determined by:

$$p(1,t) = \frac{1}{2} f_{\text{odd}}^{(2t+1)}(p)\ , \qquad p(0,t) = \frac{1}{2} - p(1,t)\ , \tag{5.52}$$

$$p(\underline{1},t) = \frac{1}{2} f_{\text{odd}}^{(2t-1)}(p)\ , \qquad p(\underline{0},t) = \frac{1}{2} - p(\underline{1},t)\ . \tag{5.53}$$

Here $f_{\text{odd}}^{(n)}(p)$ is the probability of generating an odd number of 1's from a process that generates n independent symbols (0 or 1), each with a probability p of being 1:

$$f_{\text{odd}}^{(n)}(p) = \sum_{\text{odd } k \in \{0,\ldots,n\}} \binom{n}{k} p^k (1-p)^{n-k} = \frac{1}{2}\left(1 - (1-2p)^n\right)\ . \tag{5.54}$$

In this equation, we utilised $((1-p)-p)^n = \sum_k \binom{n}{k}(-p)^k(1-p)^{n-k} = -f_{\text{odd}}^{(n)}(p) + (1 - f_{\text{odd}}^{(n)}(p))$. We see that the length of the sequence that affects a spin state at t grows linearly in time. An odd number of 1's in the original sequence at $t = 0$ results in a 1 at the position at time t. It is thus clear that, as t increases, the probability for spin up approaches 1/2, $p(\uparrow,t) = p(\underline{1},t) + p(1,t) \to 1/2$. This implies that the magnetisation approaches 0. We define the magnetisation by the difference in spin probabilities,

$$m(t) = p(\uparrow,t) - p(\downarrow,t) = 2p(\uparrow,t) - 1\ . \tag{5.55}$$

Since the frequency of state $\uparrow$ at time t is given by two binomial distributions (for updating and quiescent states, respectively) and their respective probabilities for having an odd number of 1's, we get

$$\begin{aligned} p(\uparrow,t) &= \frac{1}{2}\left(f_{\text{odd}}^{(2t-1)}(p) + f_{\text{odd}}^{(2t+1)}(p)\right) = \\ &= \frac{1}{2} - \frac{1}{4}(1-2p)^{2t-1} - \frac{1}{4}(1-2p)^{2t+1}\ , \end{aligned} \tag{5.56}$$

where we have used Eq. (5.54). Consequently, the approach towards zero magnetisation can be represented as

$$m(t) = -\frac{1}{2}\left((1-2p)^{2t-1} + (1-2p)^{2t+1}\right)\ . \tag{5.57}$$

Therefore, we observe an exponential convergence towards zero magnetisation. The frequency of spin up (and down) quickly approaches 1/2. For instance, with an initial frequency of $p = 0.1$, after $t = 10$ time steps, we get $p(\uparrow,t) = 0.4941\ldots$, and $m = -0.0118\ldots$, as illustrated in Fig. 5.6.

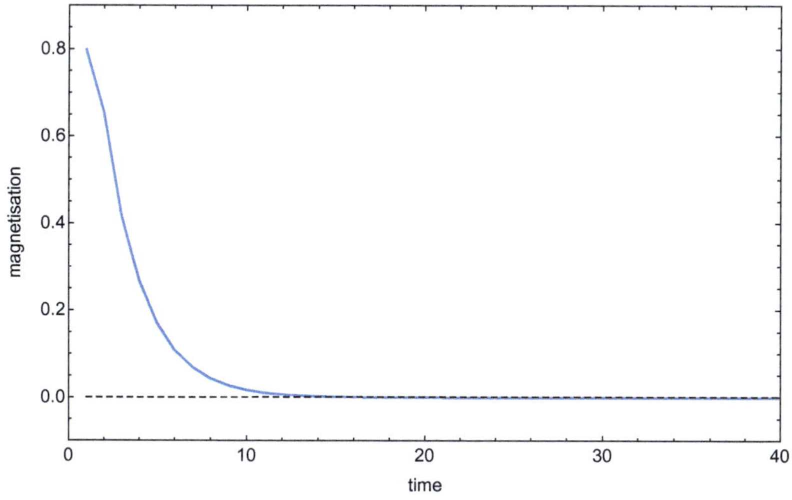

Fig. 5.6 The initial density of spin up is 0.1 and the Q2R dynamics quickly brings the spin density towards 0.5 resulting in a magnetisation approaching 0. (Since we started with dominance of spin down the magnetisation in the diagram is $-m(t)$, see Eq. (5.55))

5.6.4 Time Evolution of Information-Theoretic Characteristics

To evaluate how the Q2R dynamics transforms the initial state or the distribution of states into ones bearing similarity to equilibrium states, we employ an information-theoretic investigation of the spatial configurations at various time steps t. Essentially, we focus on the stochastic processes characterising these configurations.

The space-time pattern of the Q2R time evolution, see Fig. 5.5, clearly shows that there is a certain structure which in a sense looks relatively non-complex. It turns out, though, that the derivation of how the information characteristics change over time is non-trivial. Detailed derivations, including proofs of the propositions we present below, can be found in our previously published work (Lindgren and Olbrich 2017). These propositions primarily focus on how correlation information evolves over time. The correlation structure has some similarities with the time evolution of the simple additive (and almost reversible) elementary CA rule R60 that was discussed in Sect. 4.4.

Since the entropy s is conserved in the time evolution, so is also the total correlation information, k_{corr}, including the density information,

$$\sum_{m=1}^{\infty} k_m(t) = k_{\text{corr}} = \log 4 - s \; . \tag{5.58}$$

The total information is $\log 4$, since there are 4 possible states, and the entropy $s = S[\{p, 1-p\}]$, Eq. (5.47), where p is the initial probability for states $\uparrow$. The fixed structure of alternating quiescent and updating states means that there is always a contribution of $\log 2$ to the pair correlation information $k_2(t)$. Since there are no other correlations in the initial state, only the density information at time 0 remains, with $k_1(0) = \log 2 - s$.

A first observation we can make is that the Q2R rule, due to its local energy conservation property, conserves $p(\uparrow\uparrow) + p(\downarrow\downarrow)$ as well as $p(\uparrow\downarrow) = p(\downarrow\uparrow)$. The result derived above, that magnetisation declines towards 0, also means that $k_1(t) \to 0$ as $t \to \infty$. Therefore, in the limit of $t \to \infty$, we have that $p(\uparrow\uparrow) = p(\downarrow\downarrow) = 1/2$.

From the conservation of $p(\uparrow\uparrow) + p(\downarrow\downarrow)$ we can conclude that it always holds that $p(\uparrow\downarrow) = p(1-p)$ as it is set for the initial state. Thus, when magnetisation is 0, $p(\uparrow\uparrow) = p(\downarrow\downarrow) = (p^2 + (1-p)^2)/2$. It is clear that up to blocks of length 2, this is the equilibrium characteristics, see Sect. 5.5.1: The energy constraint is obviously fulfilled, magnetisation is 0, the entropy estimated by $S_2 - S_1$ cannot be larger due to the constraint of the given energy. In equilibrium this determines the entropy. The entropy in the Q2R system is conserved, $s = S[\{p, 1-p\}]$, while in equilibrium we have,

$$s_{\text{eq}} = S_2 - S_1 = \\ = 2p(1-p)\log\frac{1}{p(1-p)} + (p^2 + (1-p)^2)\log\frac{2}{p^2 + (1-p)^2} - \log 2\ . \tag{5.59}$$

It follows that $s_{\text{eq}} > s$ since the entropy s in the Q2R model is not maximised for the given energy. This means that as time goes on, the Q2R system approaches equilibrium characteristics for block length 2. This captures only a part of the correlation information while the remaining part is associated with longer blocks. Intuitively, this is clear from the space-time pattern shown in Fig. 5.5, where we see pairs of diagonal "lines" in two directions in space-time originating from isolated $\uparrow$ states at time 0, indicating constantly increasing correlation lengths in the system.

The propositions that follow below provide explanations for the time evolution of the correlation information. For proofs see our paper (Lindgren and Olbrich 2017).

Proposition 5.2 *There is no correlation information over blocks of even length larger than 2. For $m > 1$,*

$$k_{2m}(t) = 0\ . \tag{5.60}$$

Proposition 5.3 *A correlation information over block length $2m-1$, for $m > 1$, is in one time step transferred to a correlation information over block length $2m+1$. Thus, for $t > 0$ and $m > 1$,*

$$k_{2m+1}(t) = k_{2m-1}(t-1)\ . \tag{5.61}$$

This means that as the short range correlation information, $k_1(t)$ and $k_2(t)$, approaches equilibrium characteristics, the remaining correlation information is moving towards longer and longer blocks (of odd lengths only). This is depicted in Fig. 5.7, which is derived from the analysis presented in Lindgren and Olbrich (2017).

It is also possible to derive an expression for the time dependence of the correlation complexity, $\eta(t)$, see Eq. 3.36. This is given by the following Proposition (Lindgren and Olbrich 2017).

Proposition 5.4 *The correlation complexity, $\eta(t)$, is linearly increasing in time after the first time step.*

$$\eta(0) = \log 2 \,, \tag{5.62}$$

$$\eta(t) = (2t - 1)\zeta + \log 2 \,, \; \textit{for } t > 0, \tag{5.63}$$

where the constant ζ is determined by

$$\zeta = 2p(1-p)\log\frac{1}{2p(1-p)} + \left(p^2 + (1-p)^2\right)\log\frac{1}{p^2+(1-p)^2} - \left(p\log\frac{1}{p} + (1-p)\log\frac{1}{1-p}\right). \tag{5.64}$$

(The term $\log 2$ comes from the pair correlation information present in the periodic structure of the updating/quiescent states.) This result shows that we have an average correlation length (in the information-theoretic sense), cf. Eq. (3.37), that increases linearly in time. The immediate implication is that some of the information that was initially detectable by looking at statistics over subsequences of shorter lengths is not any longer found at those length scales.

In summary, the Q2R dynamics brings an initial non-equilibrium spin configuration with high magnetisation towards a state that has the key properties of an equilibrium state. Even though the dynamics of Q2R is reversible, and hence conserves the entropy, it seems as if entropy increases. The information-theoretic analysis reveals that the initial order, in terms of density information due to the magnetisation, is transformed into correlation information. Part of this remains at short distance: the pair correlation information that reflects the conserved energy in the system. This pair correlation information fully corresponds to the maximum entropy solution of the one-dimensional Ising model. The rest of the correlation information from the initial state is transferred to ever increasing distances, as illustrated in Fig. 5.7. In that way it becomes invisible for any finite length observations. The system appears to have an entropy that increases, as it is approaching microscopic configurations that exhibit all equilibrium properties.

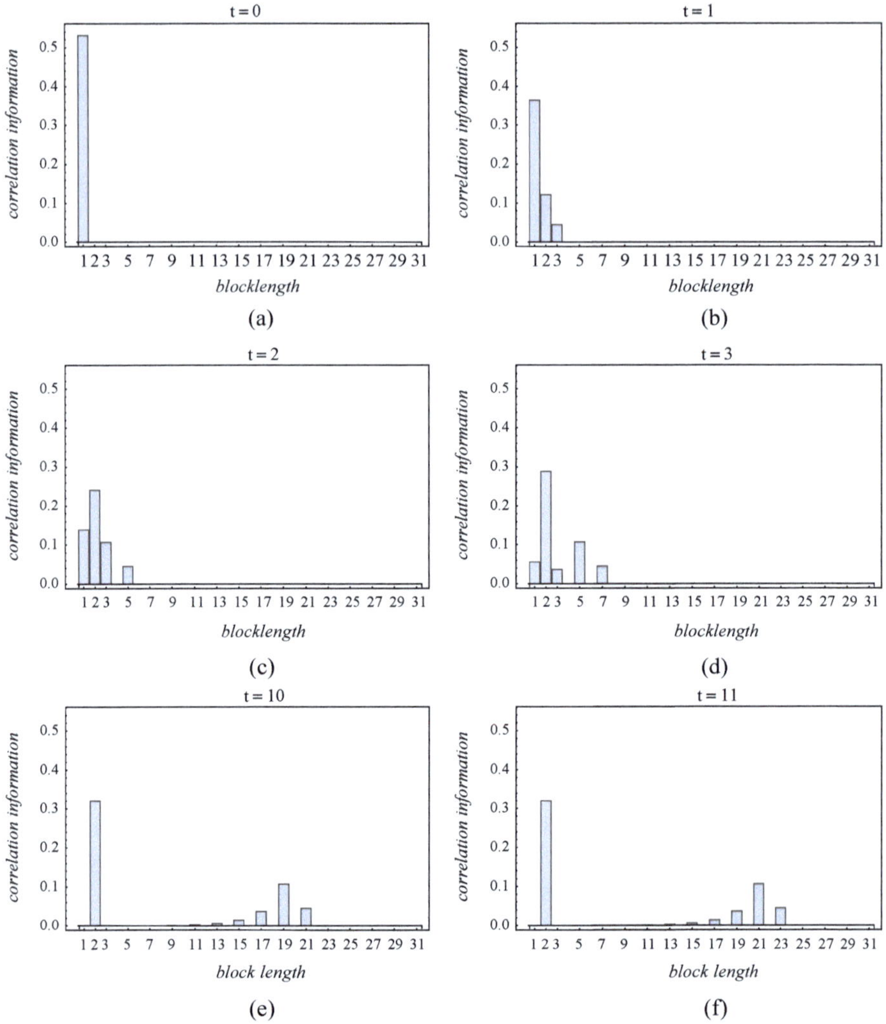

Fig. 5.7 The temporal evolution of the correlation information $k_m(t)$, for block lengths m ($1 \leq m \leq 31$), is shown for a few initial time steps $t = 0, 1, 2, 3$ and for time steps $t = 10$ and $t = 11$. The total correlation information is conserved, but the dynamics transfers some of the information to longer distances. While the density information k_1 disappears as magnetisation approaches 0, a certain part k_2 remains as pair correlation information corresponding to the conserved energy. The remaining part of the correlation information is transferred to longer and longer blocks. In this way, if one is capable of detecting only short range correlations, it appears as if correlation information is decreasing and entropy is increasing. (Here we have subtracted the constant contribution of log 2 to the pair correlation information $k_2(t)$ that comes from the regular periodic pattern of updating and quiescent states.) Figure from Lindgren and Olbrich (2017)

Problems

5.1 A small spin system

Consider a system of 2×2 spins, in which a microstate is a configuration like the one below.

$$\begin{matrix} \uparrow & \downarrow \\ \uparrow & \uparrow \end{matrix}$$

Each spin has two neighbours, as the example indicates, one in the vertical direction and one in the horizontal direction. Each such neighbour interaction contributes with an energy $+J$ for parallel spins and $-J$ for anti-parallel spins (so that in the example above the energy is 0). If the average energy is u, what is the equilibrium distribution over microstates? (Use the maximum entropy formalism. You may give the answer as a function of temperature instead of energy.)

5.2 Spin system

Suppose that in a one-dimensional discrete spin system, described by an infinite sequence of spins, up or down, the interaction is with third nearest neighbours only (position x interacts with position $x + 3$). Parallel spins at this distance contribute with an energy $-J$ and anti-parallel spins with an energy $+J$. What is the equilibrium state of this system? Use the maximum entropy formalism, and express the probabilities that characterise the equilibrium as functions of temperature (or of β).

5.3 Spin system

Consider a one-dimensional spin system with 4 states per position: $\leftarrow$, $\rightarrow$, $\downarrow$ and $\uparrow$. Interaction is with nearest neighbours only, such that parallel spins contribute with energy $-J < 0$, anti-parallel spins with the energy $J > 0$, while perpendicular spins give no energy contribution. Characterise an equilibrium distribution as function of temperature T. What is the entropy in the limit $T \rightarrow 0$? (Make use of the fact that finite interaction distance in one dimension does not allow for any phase transitions, i.e., all symmetries are kept in the equilibrium description.)

5.4 Spin system

Consider a one-dimensional spin system with 2 spin states: $\downarrow$ and $\uparrow$. Interaction is with both nearest and next nearest neighbours, so that parallel spins at distance 1 (e.g., $\uparrow\uparrow$) contribute with the energy $-J < 0$ and anti-parallel spins with energy J, while parallel spins at distance 2 (e.g., $\uparrow \_ \uparrow$) contribute with energy J and anti-parallel spins with energy $-J$. What is the equilibrium description? (The solution needs not be explicit, but it is sufficient to determine the equations from which the probabilities can be derived.) What is the entropy in the zero temperature limit ($T \rightarrow 0$), and how does a typical state look like in that situation?

5.5 Happy agents

Consider an infinite one-dimensional lattice system (of cells) in which each cell may be inhabited by a red (R) individual, a blue (B) individual, a pair (RB of agents with different colours), or the cell may be empty. Assume equal densities of individuals R and B of $1/4$ each. Each individual has a happiness level being the sum of the happiness from the relation with the closest neighbours. If there is a pair in a cell the happiness of that cell is $4H$ (with H being a positive "happiness" constant), and then there is no contribution from the neighbouring cells. If there is a single individual in a cell, the happiness of that individual get a contribution of $H/2$ from each single living neighbour of opposite colour (to the left and to the right, but no contribution from neighbour pairs). Empty cells do not contribute to happiness.

If you know that the average happiness is w, how would you guess that the system looks like in equilibrium, using information-theoretic arguments. You may answer in terms of a set of equations that you need not solve. Discuss what happens if the "temperature" is low (limit of zero temperature), with the interpretation that a low temperature corresponds to a high "happiness". What is the entropy in this limit?

(If you prefer, all this can be thought of as molecules A and B that may aggregate to a larger molecules AB, with the interpretation of H as a negative interaction energy constant.)

5.6 Spin system

Consider an infinite one-dimensional lattice system in which each position can be in one of three possible states: A, B, or C. Assume that neighbouring symbols contribute with energy $+J$ if they are the same or if they are A and C (i.e., AA, BB, CC, AC, and CA) while all other neighbour combinations contribute with $-J$, where J is a positive constant.

(a) If the average energy per position is u, what is the equilibrium distribution? Use the maximum entropy formalism to determine the equations that give the probabilistic description. You need not solve the equations derived.

(b) Calculate the entropy for zero temperature. (This can be done without solving the full problem in (a).)

5.7 Spin system automaton representation

Could the finite state automaton below possibly represent the equilibrium state in a one-dimensional spin system, in which the average internal energy u fulfils the constraint

$$u = \sum_{i_1 \ldots i_m} p(i_1 \ldots i_m) h(i_1 \ldots i_m)$$

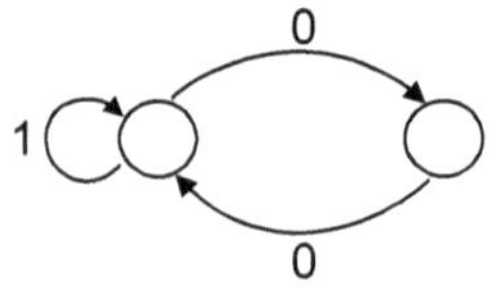

Suppose instead that the automaton that represents the equilibrium state is described by the following automaton, where q and $1 - q$ denote the transition probabilities for leaving the left node:

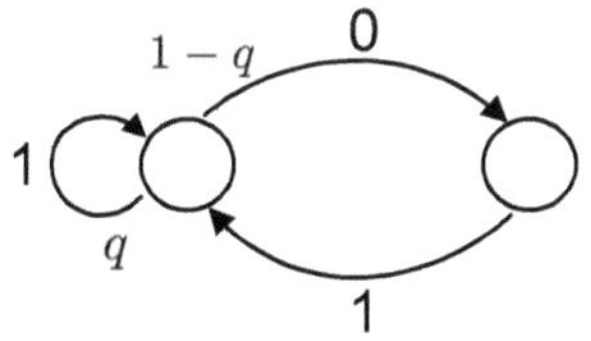

What spin system (or other one-dimensional system with energy constraints) may this automaton represent? (Must be shown.) How does the probability q connect to inverse temperature β and/or interaction energy constant J?

5.8 System of "dots"

Suppose that in a one-dimensional discrete system, described by a row of "land pieces" or cells, there live these "dot" agents. By inspecting what they are doing you quickly find the following characteristics. There is a density ρ of "dots", some of them living alone at a "piece of land", and some of them that have joined in a "marriage" so that two share the same "piece of land". You also find that whenever there is a married couple in one cell, the adjacent cells (to the left and to the right) are always free—the married couple really want to be on their own or others avoid them. You also find by studying several systems of this kind that the fraction of "dots" that are married is a certain constant α.

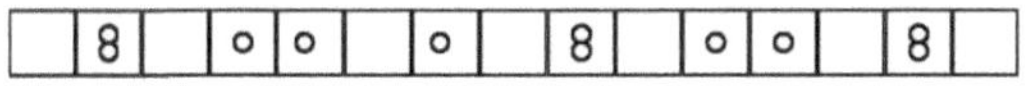

Now, before analysing more in detail the statistics of the "dot world", you want to design a probabilistic description of the system that is consistent with the observations above and that has a maximum in its (Shannon) entropy.

Use the maximum entropy formalism to find a probabilistic description that obeys the constraints described above, including the parameters ρ and α. Assume that the system is of infinite length.

5.9 Dots are back

Consider an infinite two-dimensional lattice system (of square cells) in which dots are distributed. The cells can be either empty or inhabited (by one or two dots). Free living dots have an "energy level" of 0, while pairs (two dots in one cell) have the energy $-J$ (where J is a positive constant). Assume that there is a certain density of dots ρ, and that there is a given average energy u (per cell). Use the maximum entropy formalism to characterise an "equilibrium" state of this system as a function of an "inverse temperature" (β). What is the entropy? What happens in the limit of a "zero temperature"?

Chapter 6
Geometric Information Theory

Abstract In the context of patterns or images, information theory serves as a valuable tool for identifying the length scales and positions where information is located. In this chapter, we present an information-theoretic approach to analyse patterns described by probability densities in continuous space, which can be interpreted as the light intensity representation of an image. We then decompose the relative information between a uniform a priori distribution and the observed distribution into contributions from both varying positions within the system and different length scales. This leads to an information decomposition that is different from what we presented for discrete system, and with a focus that in some cases are more applicable to images and pattens.

This chapter introduces a concept of resolution, leveraging it to establish a resolution-dependent probabilistic representation of the systems under examination. While our primary concern lies in the information decomposition with respect to position and resolution length scale, the application of resolution to observations in complex systems has a broader significance.

Resolution-dependence is a cornerstone in characterising fractal patterns and understanding the notion of 'dimension', which we will touch upon in subsequent sections. Furthermore, for dynamical systems exhibiting chaotic behaviour, observations based on limited resolution forms an essential part of the characterisation of the dynamics. This will be elaborated upon in Chap. 8. For a comprehensive overview, the reader may refer to the work by Farmer (1982).

6.1 Information Decomposition with Respect to Position and Resolution

A chemical self-organizing system may build up spatial structure in the form of concentration variations. Information-theoretic quantities for analysing such structure formation may be based on probability distributions (densities) that correspond to the spatially distributed concentrations in the system. In order to be able to analyse how

© Springer-Verlag GmbH Germany 2024

K. Lindgren, *Information Theory for Complex Systems*, Understanding Complex Systems, https://doi.org/10.1007/978-3-662-68214-2_6

information is distributed over different length scales, we introduce a "resolution" operator that can be used to modify how sharp a distribution appears. The goal is to present a formalism that can be used to decompose the total information in a pattern into contributions from both different positions and different length scales (Eriksson 1985; Eriksson and Lindgren 1987; Eriksson et al. 1987).

6.1.1 Resolution-Dependent Probability Density

In our analysis we will use a normalised probability density $p(x)$, as in Eqs. (2.34) and (2.35). We will present the formalism for a one-dimensional system, but the generalisation to higher dimensions is straightforward. Positions x are assumed to be on the real axis, $x \in \mathrm{R}$, but when applied to chemical systems, later on, we will usually assume $p(x)$ to be zero outside a finite interval (or finite volume), with periodic or non-flow boundary conditions. In case of an image, $p(x)$ can be interpreted as the light intensity of a certain colour component. Similarly, for a chemical system, $p(x)$ can be seen as the concentration profile for a certain chemical component. In both cases, the distributions can be scaled so that $p(x)$ is a normalised probability density,

$$\int_{-\infty}^{\infty} \mathrm{d}x \; p(x) = 1 \; . \tag{6.1}$$

When applied to systems of finite length L, the integral limits are 0 and L, respectively. In the following we drop the integral limits, unless they are needed for clarity. We will also need a reference, or an a priori, probability density $p_0(x)$ in order to characterise our knowledge about the system before observation. We are then assuming that observation reveals information about the system that makes us replace that a proiri distribution with a new one $p(x)$. In case of a finite system, we usually assume that the a priori description is a uniform distribution.

The amount of information that we get when we observe a spatial distribution $p(x)$ describing the spatial structure of the system is given by the relative information with the a priori distribution $p_0(x)$ as the reference,

$$K[p_0; p] = \int \mathrm{d}x \; p(x) \ln \frac{p(x)}{p_0(x)} \; . \tag{6.2}$$

We now introduce the resolution-dependent distribution, where a resolution parameter r determines how well the distribution p can be distinguished. The resolution-dependent distribution is the result of a Gaussian blur operation applied to the original one—an operation available in various types of image analysis and processing software. Mathematically, the reduced resolution is achieved by taking the convolution of the original distribution with a Gaussian of width r. The observed probability density $\tilde{p}(r; x)$ at resolution r is then written

$$\tilde{p}(r;x) = \frac{1}{\sqrt{2\pi}r}\int_{-\infty}^{\infty} \mathrm{d}w\, \mathrm{e}^{-w^2/2r^2}\, p(x-w) =$$
$$= \frac{1}{\sqrt{2\pi}}\int_{-\infty}^{\infty} \mathrm{d}z\, \mathrm{e}^{-z^2/2}\, p(x-rz)\;. \tag{6.3}$$

Parameter $r = 0$ means perfect resolution, while $r \to \infty$ leads to a Gaussian with increasing width approaching a completely uniform pattern. Any of the two expressions in Eq. (6.3) can be used to calculate the resolution-dependent distribution. For some of the theory we present, it will be useful to express the Gaussian blur operation as a differential operator. This can be achieved by making a series expansion of $p(x - rz)$ in the integrand. Note that due to symmetry, we can replace $(x - rz)$ with $(x + rz)$. Then, we can write

$$\tilde{p}(r;x) = \frac{1}{\sqrt{2\pi}}\int_{-\infty}^{\infty} \mathrm{d}z\, \mathrm{e}^{-z^2/2}\, p(x+rz) =$$
$$= \frac{1}{\sqrt{2\pi}}\int_{-\infty}^{\infty} \mathrm{d}z\, \mathrm{e}^{-z^2/2}\left(p(x) + rzp'(x) + \frac{(rz)^2}{2}p''(x) + \cdots\right) =$$
$$= \exp\left(\frac{r^2}{2}\frac{\mathrm{d}^2}{\mathrm{d}x^2}\right) p(x)\;. \tag{6.4}$$

The exponential function in the final expression should be viewed as a series expansion, involving terms of even powers of the differential operation with respect to x. By taking the derivative of the last expression with respect to r we find that the probability density fulfils

$$\left(-r\frac{\partial}{\partial r} + r^2\frac{\mathrm{d}^2}{\mathrm{d}x^2}\right)\tilde{p}(r;x) = 0\;. \tag{6.5}$$

It is also clear that the following properties hold,

$$\tilde{p}(r;x) > 0\;, \quad \text{for } r > 0\;, \tag{6.6}$$
$$\tilde{p}(0;x) = p(x)\;, \tag{6.7}$$
$$\int \mathrm{d}x\; \tilde{p}(r;x) = 1\;, \tag{6.8}$$
$$\int \mathrm{d}x\; x\,\tilde{p}(r;x) = \int \mathrm{d}x\; x\,\tilde{p}(0;x)\;, \tag{6.9}$$

where the last one states that a change in resolution r does not shift the “centre of mass” of the distribution.

6.1.2 Examples on Resolution-Dependent Probability Densities

6.1.2.1 A Gaussian Distribution

As a first example, consider a Gaussian distribution of width b,

$$p(x) = \frac{1}{\sqrt{2\pi}b} \exp\left(-\frac{x^2}{2b^2}\right) . \tag{6.10}$$

By using Eq. (6.3), we find that the resolution-dependent distribution $\tilde{p}(r; x)$ is another Gaussian, see Fig. 6.1a, with a larger width $\sqrt{b^2 + r^2}$,

$$\tilde{p}(r; x) = \frac{1}{\sqrt{2\pi}\sqrt{b^2 + r^2}} \exp\left(-\frac{x^2}{2(b^2 + r^2)}\right) . \tag{6.11}$$

6.1.2.2 Example Involving a Sinusoidal Function

For another example, consider a sinusoidal probability density,

$$p(x) = 1 + \frac{1}{2} \sin(4\pi x) , \tag{6.12}$$

on the interval $x \in [0, 1]$. One can show, using Eq. (6.3) and assuming periodic boundary conditions, that the resolution-dependent density in this case takes the form

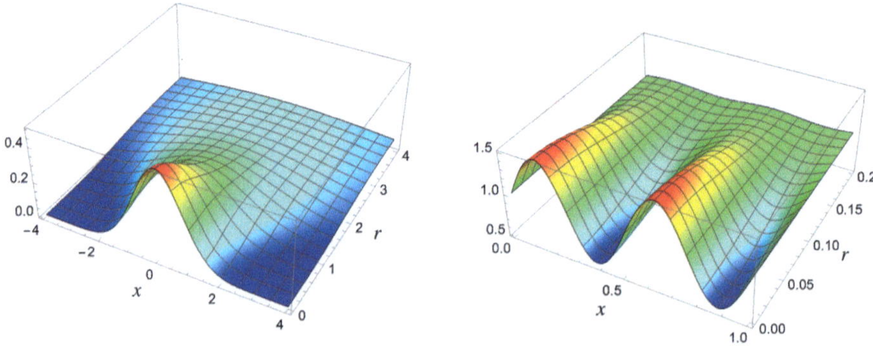

Fig. 6.1 (Left) The Gaussian probability density $\tilde{p}(r; x)$, with width $b = 0.8$, as function of position x and resolution r. (Right) A corresponding plot for the sinusoidal probability density. Periodic boundary conditions are assumed at $x = 0$ and $x = 1$

$$\tilde{p}(r; x) = 1 + \frac{1}{2}\mathrm{e}^{-8\pi^2 r^2} \sin(4\pi x) \tag{6.13}$$

The resolution-dependence of the probability density is illustrated in Fig. 6.1b.

6.1.3 Connection Between Resolution and Diffusion

A gradual increase in r, i.e., reducing the resolution or increasing the length scale at which the system is observed, can be seen as a diffusion process applied to the resolution-dependent distribution $\tilde{p}(r; x)$. If we make a variable transformation in Eq. (6.5), replacing $r^2/2$ by t, we get a diffusion equation,

$$\frac{\mathrm{d}}{\mathrm{d}t}\tilde{p}(t; x) = \frac{\mathrm{d}^2}{\mathrm{d}x^2}\tilde{p}(t; x) \; . \tag{6.14}$$

This means that the distribution $\tilde{p}(r; x)$ can be derived by running a diffusion process the time $t = r^2/2$, starting with the original distribution $p(x)$. In this perspective, Fig. 6.1 illustrates how diffusion would affect the original distributions as time goes on along the r axis.

6.1.4 Decomposition of Information

We will now consider the total information in the pattern, expressed by the relative information in Eq. (6.2), and we will derive a decomposition of this quantity into contributions from different positions x and different length scales, or resolution levels r.

First we assume that the resolution-dependent distributions $\tilde{p}(r; x)$ and $\tilde{p}_0(r; x)$ are indistinguishable in the limit $r \to \infty$,

$$\lim_{r \to \infty} \frac{\tilde{p}(r; x)}{\tilde{p}_0(r; x)} = 1 \; . \tag{6.15}$$

That this is a reasonable assumption can be seen from the fact that the ratio of any two Gaussian distributions (with centers of mass being finitely separated) always approaches 1 in the limit of worst resolution, $r \to \infty$, cf. Eq. (6.11).

We prepare for a decomposition of the relative information K, Eq. (6.2), with respect to resolution r, by subtracting from K the relative information at the infinite resolution limit $r \to \infty$, which is a zero term according to Eq. (6.15). We also use that $p(x) = \tilde{p}(0; x)$. This allows us to rewrite the relative information K as an integral over resolution r,

$$
\begin{aligned}
K[p_0; p] &= \int \mathrm{d}x\, p(x) \ln \frac{p(x)}{p_0(x)} = \\
&= \int \mathrm{d}x\, \tilde{p}(0; x) \ln \frac{\tilde{p}(0; x)}{\tilde{p}_0(0; x)} - \int \mathrm{d}x\, \tilde{p}(\infty; x) \ln \frac{\tilde{p}(\infty; x)}{\tilde{p}_0(\infty; x)} = \\
&= -\int_0^\infty \mathrm{d}r \frac{\partial}{\partial r} \int \mathrm{d}x \left(\tilde{p}(r; x) \ln \frac{\tilde{p}(r; x)}{\tilde{p}_0(r; x)} \right) = \qquad (6.16) \\
&= \int_0^\infty \mathrm{d}r \int \mathrm{d}x\, r \left[\frac{\tilde{p}(r; x)}{\tilde{p}_0(r; x)} \frac{\mathrm{d}^2}{\mathrm{d}x^2} \tilde{p}_0(r; x) - \left(1 + \ln \frac{\tilde{p}(r; x)}{\tilde{p}_0(r; x)}\right) \frac{\mathrm{d}^2}{\mathrm{d}x^2} \tilde{p}(r; x) \right] .
\end{aligned}
$$

In the last step we have used the relation between $\partial/\partial r$ and $\mathrm{d}^2/\mathrm{d}x^2$ from Eq. (6.5). By partial integration of the last expression this can be rewritten as

$$
\begin{aligned}
K[p_0; p] &= \int_0^\infty \mathrm{d}r \int \mathrm{d}x\, r\, \tilde{p}(r; x) \left(\frac{\tilde{p}'(r; x)}{\tilde{p}(r; x)} - \frac{\tilde{p}_0'(r; x)}{\tilde{p}_0(r; x)} \right)^2 = \\
&= \int_0^\infty \frac{\mathrm{d}r}{r} \int \mathrm{d}x\, \tilde{p}(r; x) \left(r \frac{\mathrm{d}}{\mathrm{d}x} \ln \frac{\tilde{p}_0(r; x)}{\tilde{p}_0(r; x)} \right)^2 , \qquad (6.17)
\end{aligned}
$$

where p' denotes derivation with respect to x. The integrand is non-negative everywhere (in x and r), and we choose to use that to define the *local information* $k(r, x)$ at position x and resolution r,

$$
k(r, x) = \tilde{p}(r; x) \left(r \frac{\mathrm{d}}{\mathrm{d}x} \ln \frac{\tilde{p}(r; x)}{\tilde{p}_0(r; x)} \right)^2 \geq 0 , \qquad (6.18)
$$

so that the total information $K[p_0; p]$ can be decomposed into local information contributions,

$$
K[p_0; p] = \int_0^\infty \frac{\mathrm{d}r}{r} \int \mathrm{d}x\, k(r, x) . \qquad (6.19)
$$

In the next chapter on chemical self-organizing systems, we will use this formalism to study the flow of information, both in space and in length scales. In the next section on fractals, we will relate resolution-dependent information quantities to the concept of fractal dimension.

In most cases we will consider a reference distribution that is uniform, at least approximately. For a limited system, i.e., a finite interval of the real axis, this is reasonable to assume. This is also the case that will be studied for chemical pattern formation in the next chapter. For an unbounded space, we typically assume p_0 to be a very broad Gaussian, which can be approximated with a uniform distribution over the region where p typically exhibits spatial pattern. In these situations, assuming a constant p_0, the local information density $k(r, x)$ can be expressed as

$$k(r, x) = r^2 p(r; x) \left(\frac{\mathrm{d}}{\mathrm{d}x} \ln p(r; x) \right)^2 . \tag{6.20}$$

From this expression, it is clear that the local information is high at the slopes of the probability density. For an image this means that information is primarily located at the edges of the objects, with edges being defined as slopes in the intensity (or probability density) $p(x)$.

For the examples discussed above, we find, using Eq. (6.20), that for the Gaussian, the information density takes the form,

$$k_{\text{Gaussian}}(r, x) = \frac{1}{(2\pi)^{1/2}(b^2 + r^2)^{5/2}} r^2 x^2 \exp\left(- \frac{x^2}{2(b^2 + r^2)} \right) , \tag{6.21}$$

and for the sinusoidal density, we get

$$k_{\sin}(r, x) = \frac{4\mathrm{e}^{-16\pi^2 r^2} \pi^2 r^2 \cos^2(4\pi x)}{1 + \frac{1}{2}\mathrm{e}^{-8\pi^2 r^2} \sin(4\pi x)} . \tag{6.22}$$

For both these examples, it is clear, see Fig. 6.2, that there are certain length scales and positions for which we have a higher information density. These length scales are then related to the width of the structure.

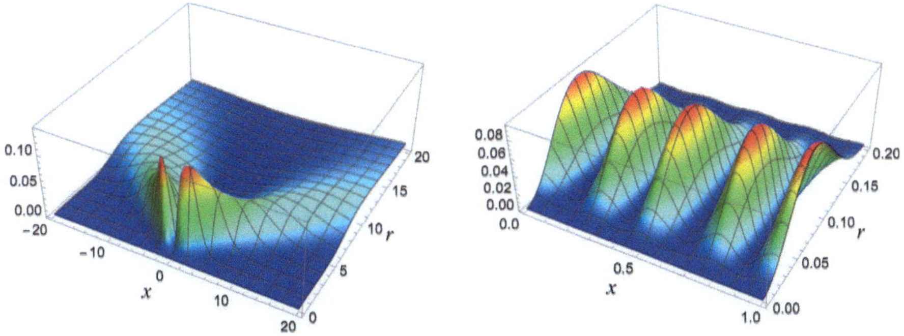

Fig. 6.2 The resolution-dependent information density $k(r, x)$ for the two examples of Fig. 6.1: (Left) The Gaussian probability density (with $b = 0.8$) is high at the edge of the blurred Gaussian and reaches its largest values around resolution level $r \approx 1$. (Right) The sinusoidal probability density is large at the slopes of the function and reaches its largest values around resolution level $r \approx 0.08$, reflecting a smaller characteristic length scale than the Gaussian

6.2 Dimension, Information, and Fractals

6.2.1 Dimensions

The word "dimension" has several definitions in mathematics. Behind the length of the coastal line of Sweden, there may be several dimension numbers hiding. A line or a curve can be considered a one-dimensional object, characterised by the fact that it can be divided into two objects by removal of a point. At the same time the object may be placed in a space of higher dimension. Furthermore, even if the coastal line is one-dimensional and constrained by a two- or three-dimensional space, it may be so irregular that its length is infinite. If a curve is strongly jagged it may be sufficiently close to any point in a two-dimensional object, and one may consider the curve two-dimensional.

The first dimension number we mentioned is related to the topology of an object—a curve may be divided by removing a point, a surface may be divided by removing a curve, etc.—and therefore we call this the *topological dimension*, D_{T}, of the object. The object is placed in a space with a certain *Euclidean* dimension, D_{E}. The third dimension number is related to how jagged, for example, a curve is. As we have indicated, this number may be larger than the topological dimension of the object, and it is actually defined so that it does not need to take only integer values. Therefore it is called the *fractal dimension*, D_{F}, of the object. An object is said to be fractal if its fractal dimension is larger than the topological dimension, $D_{\mathrm{F}} > D_{\mathrm{T}}$. In general, for these dimension numbers, we have

$$D_{\mathrm{T}} \leq D_{\mathrm{F}} \leq D_{\mathrm{E}} \ . \tag{6.23}$$

There is also another perspective on the dimension of an object, related to the resolution with which the object is observed. The following example, by Benoit Mandelbrot (1983), illustrates this:

> ... a ball of 10 cm diameter made of thick thread of 1 mm diameter possesses (in latent fashion) several effective dimensions. To an observer placed far away, the ball appears as a zero-dimensional figure: a point. (...) As seen from a distance of 10 cm resolution, the ball of thread is a three-dimensional figure. At 10 mm it is a mess of one-dimensional threads. At 0.1 mm, each thread becomes a column and the whole becomes a three-dimensional figure again. At 0.01 mm, each column dissolves into fibres, and the ball again becomes one-dimensional, and so on, with the dimensionality crossing over repeatedly from one value to another. When the ball is represented by a finite number of atom-like pinpoints, it becomes zero-dimensional again.

6.2.2 Fractal Dimension

In the following, we sketch on a definition of the fractal dimension, based on how the observed extension of the object scale when resolution is changed. The extension is related to the length of a curve, to the area of a surface, etc. Suppose that we have an object in a D_E-dimensional space. We cover the object with spheres of a certain radius r, corresponding to the resolution, in order to get a characterisation of it its extension. Let us assume that we find that $N(r)$ spheres are needed for this, where, of course, a smaller r requires a larger number of spheres, see Fig. 6.3.

If the object is a straight line the number of spheres $N(r)$ will be proportional to $1/r$, see Fig. 6.3, while in the case of a flat two-dimensional surface $N(r)$ will be proportional to $1/r^2$. Therefore, we introduce the resolution-dependent *fractal dimension*, $D(r)$, characterised by

$$N(r) \sim r^{-D(r)} . \tag{6.24}$$

We use this relation to define the fractal dimension by

$$D(r) = -\frac{\partial \ln N(r)}{\partial \ln r} . \tag{6.25}$$

In general, the fractal dimension depends on the resolution r, but in most mathematically constructed objects found in the literature on fractal patterns there is a scale invariance that makes the fractal dimension independent of the resolution, see, for example, Fig. 6.4.

The fractal dimension may also be defined by scaling of the measured length (in case of $D_T = 1$) of an object when resolution is varied. One may assume that the length scales as

$$L(r) \sim r^{1-D(r)} . \tag{6.26}$$

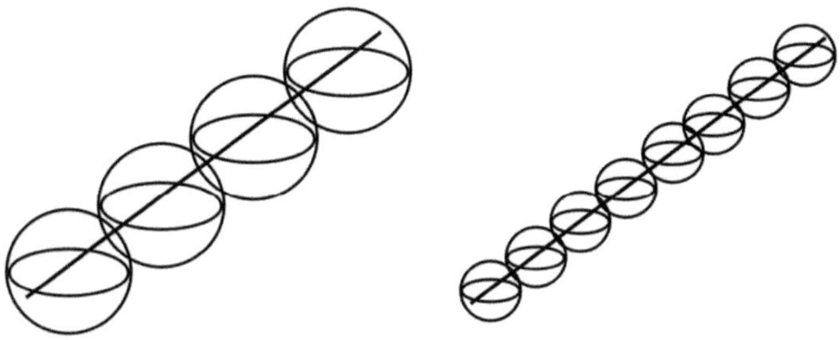

Fig. 6.3 The number of spheres with radius r that are needed to cover a curve is proportional to r^{-D}, where D is the fractal dimension of the curve. In the figure, the number of spheres doubles as the radius is reduced to half the original, implying a curve dimension $D = 1$

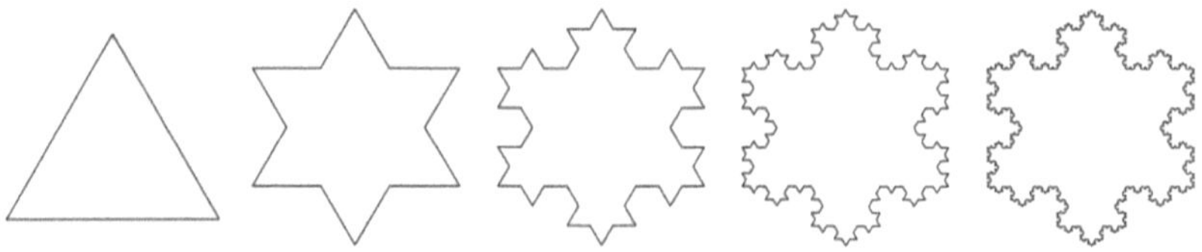

Fig. 6.4 The Koch curve is a fractal object with dimension $\ln 4/\ln 3 \approx 1.26$, in which new structures appear as soon as resolution is improved

For the West coast line of Britain one has found, within certain limits of r, that the scaling leads to a fractal dimension of $D = 1.25$ (Mandelbrot 1967), while for a straight line, the dimension is 1 since the length does not depend on the resolution (at least if the resolution is finer than the length of the object). A curve for which new jagged structures appear when resolution is improved (smaller r) may have a resolution-dependent length and a fractal dimension larger than 1.

The fractal dimension of the Koch curve in Fig. 6.4, can be derived as follows. At each step, resolution is improved by reducing r to $r/3$, i.e., an improvement by a factor of 3, and the observed length is increased from L to $4L/3$, or by a factor of $4/3$. The dimension D can then be written as

$$D(r) = 1 - \frac{\partial \ln L(r)}{\partial \ln r} = 1 - \frac{\ln L - \ln(4L/3)}{\ln r - \ln(r/3)} = \frac{\ln 4}{\ln 3} \approx 1.26 \,. \tag{6.27}$$

In this example the fractal dimension does not depend on the resolution. The object is constructed so that new structure (with the same characteristics) appears whenever resolution is improved.

6.2.3 *Dimension and Information*

If resolution is improved when we observe an object, we may gain information. Similarly, if the resolution becomes worse (r increases) we may lose information. In a three-dimensional space, a point loses information in all three directions, since we get less accurate description on three of its coordinates. For a line, on the other hand, a blurred picture only reduces information in two directions, perpendicular to the extension of the line. This indicates that there may be some connection between the dimensionality of an object and how information quantities change when resolution is varied.

Suppose that our object can be characterised by a probability density $p(\vec{x})$ in a D_{E}-dimensional Euclidean space, and further that we have an a priori probability density p_0 that is uniform. By generalisation of Eq. (6.19) to dimension D_{E}, the decomposition of the relative information with respect to both position and resolution can be written

$$K[p_0; p] = \int_0^\infty \frac{\mathrm{d}r}{r} \int \mathrm{d}\vec{x}\; p(r; \vec{x}) \left(r \frac{\mathrm{d}}{\mathrm{d}\vec{x}} \ln p(r; \vec{x}) \right)^2 . \tag{6.28}$$

The integrand thus serves as the local information density $k(r, \vec{x})$ in analogy to Eq. (6.20). By integrating the information density $k(r, \vec{x})$ over $\vec{x}$, we get the information $k(r)$ at length scale (resolution) r,

$$k(r) = \int \mathrm{d}\vec{x}\; \tilde{p}(r; \vec{x}) \left(r \frac{\mathrm{d}}{\mathrm{d}\vec{x}} \ln \tilde{p}(r; \vec{x}) \right)^2 . \tag{6.29}$$

The information $k(r)$ at a certain level of resolution can be related to the change in resolution-dependent entropy,

$$k(r) = r \frac{\partial}{\partial r} S[\tilde{p}] , \tag{6.30}$$

where the entropy is given by

$$S[\tilde{p}] = \int \mathrm{d}\vec{x}\; \tilde{p}(r; \vec{x}) \ln \frac{1}{\tilde{p}(r; \vec{x})} . \tag{6.31}$$

This is derived by carrying out the derivation of S with respect to r, using the relation between $\partial/\partial r$ and $\mathrm{d}^2/\mathrm{d}x^2$ from Eq. (6.5),

$$\begin{aligned} r \frac{\partial}{\partial r} S[\tilde{p}] &= -r^2 \int \mathrm{d}\vec{x}\; \frac{\mathrm{d}^2}{\mathrm{d}\vec{x}^2} \tilde{p}(r; \vec{x})(1 + \ln \tilde{p}(r; \vec{x})) = \\ &= r^2 \int \mathrm{d}\vec{x}\; \left(\frac{\mathrm{d}}{\mathrm{d}\vec{x}} \tilde{p}(r; \vec{x}) \right)^2 \frac{1}{\tilde{p}(r; \vec{x})} = \qquad (6.32) \\ &= \int \mathrm{d}\vec{x}\; k(r, \vec{x}) = k(r) , \qquad (6.33) \end{aligned}$$

where we have also used partial integration, assuming $\mathrm{d}\tilde{p}/\mathrm{d}\vec{x}\ (1 + \ln \tilde{p}) \to 0$ when $|\vec{x}| \to \infty$.

We now *define* the dimension $d(r)$ as an information-theoretic counterpart to $D(r)$, following the discussion above that a low-dimensional object loses more information than a higher-dimensional one when resolution becomes worse,

$$d(r) = D_\mathrm{E} - r \frac{\partial}{\partial r} S[\tilde{p}] = D_\mathrm{E} - k(r) . \tag{6.34}$$

As a schematic illustration on how this dimension quantity is related to the fractal dimension defined in Eq. (6.25), we approximate the probability density $p(r; \vec{x})$ with a uniform distribution over the $N(r)$ spheres covering the object as it was illustrated in Fig. 6.3, i.e.,

$$\tilde{p}(r;\vec{x}) \sim \frac{1}{N(r)r^{D_{\mathrm{E}}}} . \tag{6.35}$$

This means that the entropy of $\tilde{p}$ is approximated by

$$S[\tilde{p}] = \int \mathrm{d}\vec{x}\; \tilde{p}(r;\vec{x}) \ln(CN(r)r^{D_{\mathrm{E}}}) , \tag{6.36}$$

where C is a constant. Then, using Eq. (6.34) the dimension $d(r)$ can be derived,

$$d(r) = D_{\mathrm{E}} - r\frac{\partial}{\partial r} \ln\left(CN(r)r^{D_{\mathrm{E}}}\right) = -\frac{\partial \ln(N(r))}{\partial \ln r} = D(r) , \tag{6.37}$$

where we in the last step used the definition of fractal dimension, Eq. (6.25). This illustrates the connection between information-theoretic dimension $d(r)$ and fractal dimension $D(r)$.

Problems

6.1 Dimension of a Gaussian
What is the dimension $d(r)$ of a Gaussian distribution of width b (variance b^2) in one dimension ($D_{\mathrm{E}} = 1$)? How can this be interpreted? Discuss the relation between r and b.

6.2 Dimension of a ring
Consider a pattern in a two-dimensional space that is described as a ring with a certain thickness a and a certain diameter $10a$, schematically depicted in the figure below.

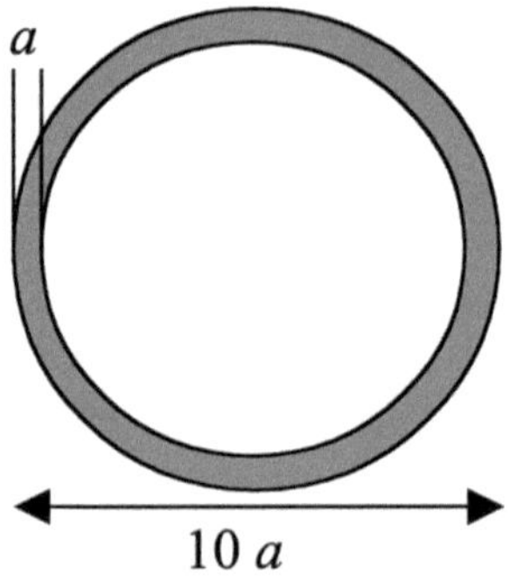

The ring may be described by a uniform probability distribution $p(x, y)$ that is constant within the grey area and zero elsewhere. By introducing the resolution-dependent probability density $p(r; x, y)$, one may study how the entropy $S(r)$ changes when the resolution is made worse (r increases).

Discuss how the quantity

$$r\frac{\partial}{\partial r}\int \mathrm{d}x\,\mathrm{d}y\; p(r;x,y)\ln\frac{1}{p(r;x,y)}$$

depends on r. It should not be necessary with calculations (which are quite complicated), but you should be able to discuss this in a more qualitative way, relating resolution level r to characteristic lengths in the system, for example a.

6.3 Gaussians indistinguishable at worst resolution
What is the relative information between two Gaussian distributions (with centres of mass being finitely separated) in the limit of worst resolution, $r \to \infty$, cf. Eq. (6.11).

Chapter 7
Pattern Formation in Chemical Systems

Abstract The spontaneous formation of patterns in chemical systems is a phenomenon that is interesting from both the information-theoretic and the thermodynamic perspective. How can a seemingly increase in order, when a spatial pattern is formed, be explained with the second law of thermodynamics as a force that brings systems closer to equilibrium? How can spatial chemical patterns be characterised in terms of information? How can this be related to the thermodynamics of the chemical system? The aim of this chapter is to answer these question, using the information decomposition along with the thermodynamics perspective, both presented in the previous chapters. With this approach, we can derive a continuity equation for information that is determined by the reaction-diffusion dynamics of the chemical system. Due to the connection between the information in a chemical system and the chemical free energy it represents, we can connect the physical or thermodynamic constraints of the pattern formation process to the information characteristics of the system. The formalism is applied to the self-replicating spots system based on the Gray-Scott reaction-diffusion dynamics.

In this chapter we apply information theory to systems exhibiting chemical pattern formation, often called chemical self-organizing systems. Here self-organization refers to a system that spontaneously builds up or sustain spatio-temporal structure in the form of concentration variations. The patterns formed, and the associated chemical dynamics, is not controlled by anything outside the system. The first theoretical investigation of chemical pattern formation was done by Alan Turing in his classic paper, where he demonstrated that simple mechanisms in reaction-diffusion dynamics can account for symmetry breaking necessary for morphogenesis (Turing 1952). Self-organizing systems, or dissipative structures, that also go beyond chemical pattern formation have been extensively studied, see, for example the work by Nicolis and Prigogine (1977) and Haken (1984).

There are several types of information quantities involved in pattern formation processes, and they are typically of different orders of magnitude. A first piece of information when one is presented with a specific chemical system is the selection of molecules involved, both those that are present in the system and those that are

© Springer-Verlag GmbH Germany 2024
K. Lindgren, *Information Theory for Complex Systems*, Understanding Complex Systems, https://doi.org/10.1007/978-3-662-68214-2_7

allowed to flow over the system border. All this can loosely be referred to as genetic information, i.e., information on which components and thus also which processes that will be part of the self-organizing chemical system. The amount of this information is not very large, as is reflected for example by the size of genomes in living organisms with the order of 10^4–10^5 genes.

In biological systems all necessary information is not genetically encoded, but there is also compositional information in the transfer of chemicals and structures from the mother to the daughter cells. There are proposals expressing the idea that this type of information may have played an important role in the origin of life (Segré et al. 2000).

Another type of information enters when a specific self-organizing system starts to develop a pattern. The typical form of the pattern may be determined by the reaction scheme involved, but in many cases fluctuations in concentrations or other disturbances may affect the exact pattern that is formed. An example of that is the difference in finger prints between identical twins. One may view this as an information flow from fluctuations to the actual pattern that is observed. This flow is of the same character as the flow from micro to macro that we find in chaotic systems (Shaw 1981). This information flow can be characterised by the Lyapunov exponent, an important quantity for the analysis of chaotic dynamical systems. This perspective will be brought up in the next chapter presenting an information-theoretic perspective on low-dimensional chaotic systems.

The focus of the approach presented here is a third type of information quantity, the information capacity in free energy or exergy, based on the information-theoretic formalism presented in Chap. 5. This is then combined with the geometric information theory of Chap. 6. The starting point is the exergy (or the free energy) of a chemical system, involving both the deviation from homogeneity when a spatial pattern is present and a deviation from chemical equilibrium (if the stirred system would still deviate from equilibrium). This exergy is expressed as the total information of the system, and in our approach we decompose this into information contributions from both different positions and different length scales. In case of an open system, exchanging fuel and waste products with an environment, the connection with thermodynamics then allows us to view the inflow of exergy (chemical free energy) as an inflow of information capacity. This inflow allows for an accumulation of information in the system when a pattern is formed. Entropy production due to chemical reactions and diffusion leads to destruction of information—an information loss that can be balanced by the information capacity inflow to maintain the chemical pattern. The following presentation is based on our previously published work, see Eriksson and Lindgren (1987), Eriksson et al. (1987), Lindgren et al. (2004) and Lindgren (2015).

7.1 The Transition from Micro to Macro

Before presenting the analytical framework for analysis of patterns in chemical reaction-diffusion systems, a short discussion on the difference between a microscopic and a macroscopic description will serve as a background. This is a conceptual discussion aiming for clarification of some relations between information-theoretic and thermodynamic system characteristics in the context of microscopic and macroscopic descriptions.

In Chap. 5 on physics and information theory, a comprehensive microscopic portrayal of dynamics in a physical spin system was outlined, highlighting the dichotomy of order and disorder and its relation to thermodynamic entropy. In principle, it is possible to model also macroscopic pattern formation in chemical systems on the microscopic level. However, such an approach would require an enormous number of molecules, both in the model and in the simulation. It is also well known that chemical systems, developing in time through chemical reactions and diffusion, can be very well represented on the macroscopic level employing concentrations as variables and reaction-diffusion partial differential equations to represent the dynamics. Nevertheless, in order to fully understand the information characteristics of the macroscopic description, it is necessary to see how that is related to microscopic information quantities.

As a theoretical exercise, let us consider a deterministic lattice gas scenario. Imagine a triangular lattice populated with up to six particles at each lattice site, comprising two particle types (A and B). Each particle possesses a unit velocity, pointing in one of the six possible lattice directions. This implies that we have 3^6 possible states for each lattice point. A microscopic dynamics can be formulated as follows. A time step consists of two stages: (i) Collision rules apply when particles collide, ensuring energy and momentum conservation. These rules can be designed to be reversible and deterministic. (ii) Particles then migrate to neighbouring lattice points as guided by their velocity. A visual depiction of a microstate after collisions and before translation can be found in Fig. 7.1. This means that we have a perfectly reversible microscopic dynamics, where the entropy per lattice site remains constant over time. This model can be viewed as a deterministic two-component version of the FHP lattice gas, as proposed by Frisch, Hasslacher, and Pomeau (Pomeau et al. 1986).

In order to discuss the transition from micro to macro, let us now consider a microstate of a vast system composed of many particles. In this scenario, the concentration variations extend across vast spatial scales, and we are assuming a relatively low density, hence the majority of lattice sites remain vacant. Given a randomly dispersed initial state of A and B particles, arranged as concentration peaks of particle A against a background of uniformly distributed B particles, we attain a state with lower entropy per lattice site, s, compared to a situation where *all* particles were distributed evenly.

The lower entropy of this initial state corresponds to information that captures deviation from equilibrium, manifested as a higher density of A particles around

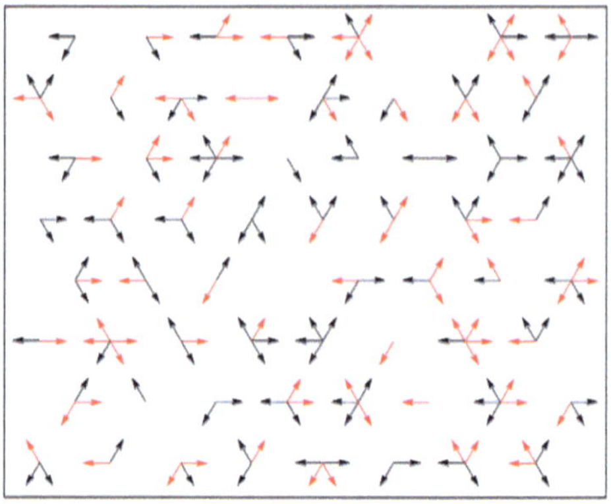

Fig. 7.1 An illustration of the state in a two-component lattice gas on a triangular lattice (with black and red particles), in which particles have collided and are about to take the translational step moving to adjacent cells. Figure from Lindgren (2015)

concentration peaks. This holds both for the microscopic representation in the FHP lattice model and for the macroscopic description using continuous concentration variables.

As the micro-dynamics unfolds, particles move and collide reversibly, resulting in the formation of longer and longer correlations of increasing order. This process mimics diffusion, evidenced by the gradual disappearance of the initial concentration peaks. This phenomenon is consistent within both the microscopic and macroscopic descriptions. However, the full description on the micro-level, with its reversible dynamics, retains the complete information of the initial state, thus maintaining constant microscopic entropy. The micro-dynamics bear a similarity to the reversible Ising dynamics model Q2R discussed in Sect. 5.6. The discussion regarding the approach towards equilibrium, in terms of time evolution at the microscopic level, bears significant relevance to both the spin system dynamics and the lattice gas model.

The information that is, in general, available at the *macroscopic* level, as a result of measurement of the system, is in the form of the local concentration, $c_i(\vec{x})$ for chemical species i at position $\vec{x}$. This does not include exact positions and velocities of individual particles, hence the correlation information that is built up in the micro dynamics is absent in the macroscopic description. At the micro-level, we have a representation capturing the full physical description of the system, including long-range correlations.

Our theoretical experiment corresponds to the mixing of ideal gases during diffusion. The macro-level diffusion process can be expressed as,

$$\frac{dc_i(\vec{x})}{dt} = D_i \, \nabla^2 c_i(\vec{x}) \; , \tag{7.1}$$

where D_i is the diffusion constant for molecular species i. In this very specific theoretical experiment, where the particles are moving with unit velocities on a lattice, the diffusion constant is the same for all i.

The change in macroscopic entropy S, can then be obtained by

$$\frac{dS}{dt} = \frac{d}{dt}\int_V d\vec{x} \sum_i c_i(\vec{x},t)\log\frac{1}{c_i(\vec{x},t)}$$
$$= \int_V d\vec{x} \sum_i D_i \frac{(\nabla c_i(\vec{x},t))^2}{c_i(\vec{x},t)} \geq 0\,, \tag{7.2}$$

where we used partial integration, assuming periodic or non-flow boundary conditions. This is the standard illustration that the total entropy of a closed system cannot decrease under diffusion.

The entropy increase, as seen in the macroscopic description relative to the conserved entropy at the microscopic level, hinges upon the fact that on the macro level, information is lost due to increasing long-range high-order correlations. This increase in macroscopic entropy, paired with a corresponding decrease in macroscopic information (order), can be interpreted as an information flow from the macroscopic scales dissipating into microscopic degrees of freedom as correlation information. This understanding will serve as a useful conceptual tool when we present the information dynamics framework for chemical pattern formation in the following sections.

7.2 Information Analysis of Chemical Pattern Formation

A chemical self-organizing system is driven by the exergy that is present in the form of chemical free energy. We assume that the system has the same pressure and the same temperature as the environment. According to Eq. (5.37), the exergy for a homogenous system can be written as a relative information

$$E = Vk_\mathrm{B}T_0\sum_{i=1}^{M} c_i \ln\frac{c_i}{c_{i0}} = Nk_\mathrm{B}T_0 K[P_0;\,P]\,, \tag{7.3}$$

where the distributions P_0 and P denote the normalised distributions $p_{i0} = Vc_{i0}/N$ and $p_i = Vc_i/N$, respectively.

The reference system P_0 represents a system in reaction equilibrium, and the non-equilibrium of the present system, characterised by P, can be arranged in two ways. In case of a closed system, P is initially representing how the system is prepared with chemical components out-of-equilibrium, typically an abundance of a “fuel” component that may be exploited in the process of pattern formation. Over time, the distribution P changes as the fuel is consumed and equilibrium is approached. In the case of an open system where chemical substances are allowed to pass the system border, the concentrations in the system may be kept off equilibrium so that exergy is available both for the creation and the maintenance of spatial patterns.

7.2.1 Chemical and Spatial Information

We shall use Eq. (7.3) as a starting point for our combined information-theoretic and thermodynamic analysis of the system. Since homogeneity is assumed to be broken, we introduce spatially varying concentrations $c_i(\vec{x})$ assumed to be normalised in each position $\vec{x}$. Thus $c_i(\vec{x})$ serve as probability distributions over the different molecular species i that are normalised at each position,

$$\sum_{i=1}^{M} c_i(\vec{x}) = 1 \, . \tag{7.4}$$

We also use normalised concentrations c_{i0} characterising the equilibrium situation. Then, with this modification in notation, Eq. (7.3) is replaced by

$$E = k_B T_0 \frac{N}{V} K \, , \tag{7.5}$$

where the information K is now an integral over relative information quantities for each position in the system,

$$K = \int_V \mathrm{d}\vec{x} \sum_{i=1}^{M} c_i(\vec{x}) \ln \frac{c_i(\vec{x})}{c_{i0}} \, . \tag{7.6}$$

We shall use the average concentration within the system, defined by

$$\overline{c}_i = \frac{1}{V} \int_V \mathrm{d}\vec{x} \; c_i(\vec{x}) \, , \tag{7.7}$$

in order to decompose the total information K in Eq. (7.6) into two terms, both in the form of relative information. One term quantifies the deviation of the average concentrations from equilibrium. The second term quantifies the deviation from homogeneity, i.e., the presence of spatial structure,

$$\begin{aligned} K &= \int_V \mathrm{d}\vec{x} \sum_{i=1}^{M} c_i(\vec{x}) \ln \frac{c_i(\vec{x})}{\overline{c}_i} \frac{\overline{c}_i}{p_{i0}} = \\ &= \int_V \mathrm{d}\vec{x} \sum_{i=1}^{M} c_i(\vec{x}) \ln \frac{c_i(\vec{x})}{\overline{c}_i} + V \sum_{i=1}^{M} \overline{c}_i \ln \frac{\overline{c}_i}{c_{i0}} \, . \end{aligned} \tag{7.8}$$

Consequently, we define the spatial information, K_{spatial},

$$K_{\text{spatial}} = \int_V \mathrm{d}\vec{x} \sum_{i=1}^{M} c_i(\vec{x}) \ln \frac{c_i(\vec{x})}{\overline{c}_i} \geq 0 \, , \tag{7.9}$$

and the chemical information, K_{chem},

$$K_{\text{chem}} = V \sum_{i=1}^{M} \overline{c}_i \ln \frac{\overline{c}_i}{c_{i0}} \geq 0 \,. \tag{7.10}$$

Thus we can write the total information K as

$$K = K_{\text{spatial}} + K_{\text{chem}} \,. \tag{7.11}$$

Thermodynamically, the chemical information is related to the presence of chemical non-equilibrium even when spatial variations are not taken into account. This means that there is an abundance of "fuel" and a low level of "waste" products in the system. The spatial information reflects the presence of non-equilibrium for diffusion processes, i.e., that there is some spatial pattern in the system.

7.2.2 Decomposition of Spatial Information in a Chemical Pattern

We continue our analysis by a further decomposition of the spatial information into contributions from different positions and length scales, similar to the approach in Chap. 6.

In order to be able to analyse contributions from different length scales, we introduce a resolution-dependent concentration $\tilde{c}_i$, cf. Eq. (6.3), by convolution of the original one c_i with a Gaussian,

$$\tilde{c}_i(r, \vec{x}) = \frac{1}{(2\pi)^{n/2}} \int d\vec{z}\; d^{-z^2/2} c_i(\vec{x} + r\vec{z}) = \tag{7.12}$$

$$= \exp\left(\frac{r^2}{2}\nabla^2\right) c_i(\vec{x}) = R(r) c_i(\vec{x}) \,. \tag{7.13}$$

We will use the last expression, Eq. (7.13), as a resolution operator applied to the concentration for analytic derivations, but when calculating $\tilde{c}_i$, the first expression, Eq. (7.12), will be used. We assume that this operation handles the boundary conditions so that in the limit of $r \to \infty$, i.e., complete loss of position information, the concentrations equal the average concentrations in the system,[1]

$$\tilde{c}_i(\infty, \vec{x}) = \overline{c}_i \,. \tag{7.14}$$

[1] In the simple examples we study, this follows immediately from the assumption on periodic boundary conditions.

The derivation of a decomposition of the total information K into different contributions will now be slightly different from the one in Chap. 6, since the distributions are now normalised in each position (instead of over the system volume). Our starting point, in Chap. 6, for decomposing the total information K, is used again,

$$\int_0^\infty \mathrm{d}r \, \frac{\partial}{\partial r} \int \mathrm{d}\vec{x} \; K[c_0; \tilde{c}(r, \vec{x})] = \int \mathrm{d}\vec{x} \; K[c_0; \overline{c}] - \int \mathrm{d}\vec{x} \; K[c_0; \tilde{c}(0, \vec{x})]$$
$$= K_{\mathrm{chem}} - K \; , \tag{7.15}$$

where we have used the fact that $\tilde{c}(0, \vec{x}) = c(\vec{x})$. This illustrates that the chemical information K_{chem} can be detected regardless of how bad the resolution is. This is obvious, since complete loss of resolution results in average concentrations in the system which determine the chemical information. This means that the spatial information K_{spatial} can be decomposed in a similar way as in Chap. 6,

$$\begin{aligned}
K_{\mathrm{spatial}} = K - K_{\mathrm{chem}} &= -\int_0^\infty \mathrm{d}r \frac{\partial}{\partial r} \int \mathrm{d}\vec{x} \; K[c_0; \tilde{c}(r, \vec{x})] = \\
&= -\int_0^\infty \mathrm{d}r \int \mathrm{d}\vec{x} \; \frac{\partial}{\partial r} \sum_i \tilde{c}_i(r, \vec{x}) \ln \frac{\tilde{c}_i(r, \vec{x})}{c_{i0}} = \\
&= -\int_0^\infty \mathrm{d}r \int \mathrm{d}\vec{x} \; \sum_i \frac{\partial \tilde{c}_i(r, \vec{x})}{\partial r} \ln \left(\frac{\tilde{c}_i(r, \vec{x})}{c_{i0}} + 1 \right) = \\
&= -\int_0^\infty \mathrm{d}r \int \mathrm{d}\vec{x} \; \sum_i r \nabla^2 \tilde{c}_i(r, \vec{x}) \ln \frac{\tilde{c}_i(r, \vec{x})}{c_{i0}} = \\
&= \int_0^\infty \frac{\mathrm{d}r}{r} \int \mathrm{d}\vec{x} \; \sum_i \tilde{c}_i(r, \vec{x}) \left(r \nabla \ln \frac{\tilde{c}_i(r, \vec{x})}{c_{i0}} \right)^2 .
\end{aligned} \tag{7.16}$$

In the last step, a partial integration is involved, including an assumption of periodic (or non-flow) boundary conditions. (We also used the fact that $\sum_i \partial \tilde{c}_i / \partial r = 0$ due to normalisation.) The equilibrium concentration c_{i0} in the last expressions could be removed (as it disappears under the differential operation), but we keep it as it will serve a purpose in later derivations. In the last expression we recognise a local information density, $k(r, \vec{x})$, which can be written in three different forms, of which the last one will be used later,

$$k(r, \vec{x}) = r^2 \sum_i \tilde{c}_i(r, \vec{x}) \left(\nabla \ln \frac{\tilde{c}_i(r, \vec{x})}{c_{i0}} \right)^2 = \tag{7.17}$$

$$= r^2 \sum_i \frac{(\nabla \tilde{c}_i(r, \vec{x}))^2}{\tilde{c}_i(r, \vec{x})} = \tag{7.18}$$

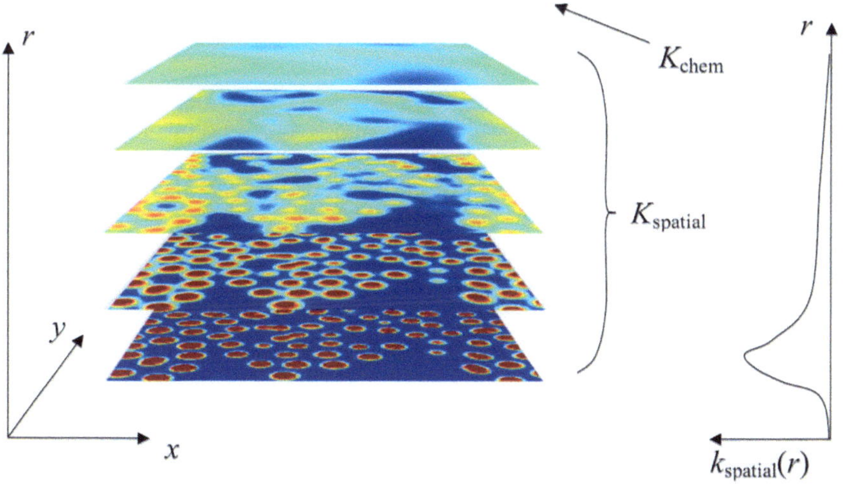

Fig. 7.2 An illustration of a two-dimensional chemical pattern at different levels of resolution r, along with a schematic illustration on how the spatial information may be decomposed between different length scales. The chemical information can be detected regardless of how bad the resolution is, and it is therefore placed at infinite r. Figure from Lindgren (2015)

$$= \left(-r\frac{\partial}{\partial r} + r^2\nabla^2\right)\sum_i \tilde{c}_i(r,\vec{x})\ln\frac{\tilde{c}_i(r,\vec{x})}{c_{i0}} . \tag{7.19}$$

The information density $k(r, x)$ can be integrated over space so that we achieve the spatial information at a certain length scale,

$$k_{\text{spatial}}(r) = \int d\vec{x}\; k(r,\vec{x}) . \tag{7.20}$$

A schematic illustration of this decomposition is shown in Fig. 7.2, illustrating a chemical pattern in a two-dimensional space with its variations along the axis of resolution r.

7.2.3 Reaction-Diffusion Dynamics

The concentrations are time-dependent since they vary in time due to diffusion, characterised by diffusion constants D_i, and chemical reactions, characterised by reaction functions $F_i(\vec{c}(\vec{x}, t))$, where $\vec{c} = (c_1, \ldots, c_M)$. The equations of motion governing the dynamics is then the ordinary reaction-diffusion equations plus a term $B_i(c_i(\vec{x}, t))$ capturing the flow of chemical component i across the system border in the case of an open system,

$$\frac{\mathrm{d}}{\mathrm{d}t}c_i(\vec{x},t) = \dot{c}_i(\vec{x},t) = D_i\nabla^2 c_i(\vec{x},t) + F_i(\vec{c}(\vec{x},t)) + B_i(c_i(\vec{x},t)) \;, \tag{7.21}$$

for the different chemical species, $i = 1, \ldots, M$. The term $B_i(c_i(\vec{x},t))$ is assumed to be in the form of diffusion-controlled flows, as detailed below. We assume that the in- and out-flow is directly connected to the whole system. For example, in a two-dimensional system, this means that there is a flow across the "surface" of the system in the direction of a third dimension. We can view this as if the reaction volume everywhere is in contact with a reservoir having a constant concentration $c_i^{(\mathrm{res})}$ which results in an inflow

$$B_i(c_i(\vec{x},t)) = b_i\left(c_i^{(\mathrm{res})} - c_i(\vec{x},t)\right) \;. \tag{7.22}$$

with b_i being a diffusion constant. A negative value of that expression reflects an outflow. The open reaction-diffusion system, driven by the reservoir, is illustrated in Fig. 7.3. The a priori, or reference, concentrations, c_{i0}, represent an equilibrium system, but the chemical dynamics is only directly driven by the in- and outflow associated with the reservoir. The reservoir is characterised by concentration levels, $c_i^{(\mathrm{res})}$, that are assumed to be kept constant.

The equations of motion for the resolution-dependent concentrations are derived from Eq. (7.21) by applying the resolution operator, see Eq. (7.12), to both sides of the equation,

$$\dot{\tilde{c}}(r,\vec{x},t) = D_i\nabla^2\tilde{c}_i(r,\vec{x},t) + \tilde{F}_i(\vec{c}(\vec{x},t)) + \tilde{B}_i(c_i(\vec{x},t)) \;, \tag{7.23}$$

where we have used the fact that the resolution operator $(\exp(r^2\nabla^2/2))$, or equivalently the Gaussian convolution, commutes with the diffusion operator (∇^2). The terms $\tilde{F}_i$ and $\tilde{B}_i$ represent the result of the convolution applied to the chemical reaction term and the flow term, respectively. In general, those terms cannot be simplified, but if the flow is linearly dependent on the concentrations, as in Eq. (7.22), we can replace c_i by $\tilde{c}_i$ in the last term,

$$\dot{\tilde{c}}(r,\vec{x},t) = D_i\nabla^2\tilde{c}_i(r,\vec{x},t) + \tilde{F}_i(\vec{c}(\vec{x},t)) + b_i\left(c_i^{(\mathrm{res})} - \tilde{c}_i(\vec{x},t)\right) \;. \tag{7.24}$$

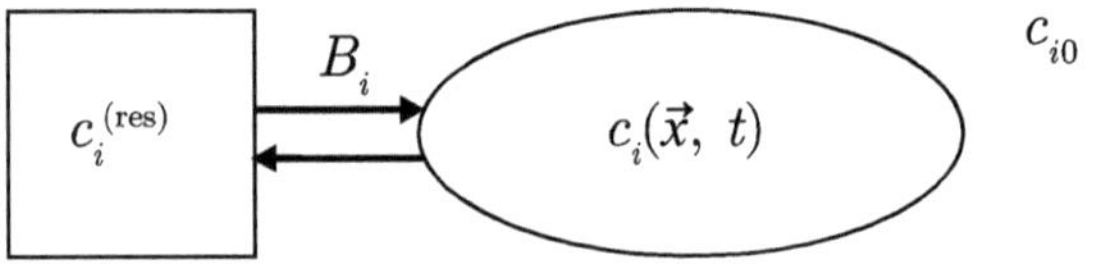

Fig. 7.3 A schematic illustration of the open reaction-diffusion system, with concentrations $c_i(\vec{x},t)$. There is a diffusion-controlled flow B_i, between the system and a reservoir, in which the concentrations, $c_i^{(\mathrm{res})}$, are assumed to be kept constant. As a reference, the environment is assumed to be in homogenous equilibrium characterised by concentrations c_{i0}.

This equation will be used in the derivation of suitable quantities for information flows in reaction-diffusion dynamics.

7.2.4 Destruction of Information in Closed Chemical Systems

In a closed homogenous chemical system, prepared in an out-of-equilibrium state, all information is initially in the form of chemical information, K_{chem}, corresponding to an initial amount of exergy. Chemical reactions consume the exergy, according to the 2nd law of thermodynamics, and the chemical information thus decays. If spatial structure is built up, we find that some of the information, at least temporarily, is transformed to spatial information, K_{spatial}. If the system remains closed, though, such spatial structures cannot remain and the system eventually approaches a homogenous equilibrium state. In an open system, a steady inflow of chemical free energy may keep the chemical information at a high level, and spatial structures may be supported. First we will discuss the closed system and derive a set of information flow quantities, and in the next section the system will be open for molecular flows across the boundary which will result in additional effects. So to start with, we neglect the flow term B in Eqs. (7.21) and (7.24).

The reaction-diffusion dynamics changes the chemical composition in the system. A first question is how this dynamics affect the total information $K = K(t)$ of the system, or equivalently the exergy of the system. By using the dynamics of Eq. (7.21), with $B = 0$, the time derivative of the total information can be written

$$\frac{\mathrm{d}}{\mathrm{d}t}K(t) = \frac{\mathrm{d}}{\mathrm{d}t}\int_V \mathrm{d}\vec{x} \sum_i c_i(\vec{x},t) \ln \frac{c_i(\vec{x},t)}{c_{i0}} = \int_V \mathrm{d}\vec{x} \sum_i \dot{c}_i(\vec{x},t)\Big(\ln \frac{c_i(\vec{x},t)}{c_{i0}} + 1\Big) =$$
$$= \int_V \mathrm{d}\vec{x} \sum_i \Big(D_i \nabla^2 c_i(\vec{x},t) + F_i(\vec{c}(\vec{x},t))\Big) \ln \frac{c_i(\vec{x},t)}{c_{i0}} =$$
$$= \int_V \mathrm{d}\vec{x} \Big(- \sum_i D_i \frac{(\nabla c_i(\vec{x},t))^2}{c_i(\vec{x},t)} + \sum_i F_i(\vec{c}(\vec{x},t)) \ln \frac{c_i(\vec{x},t)}{c_{i0}} \Big). \tag{7.25}$$

Here we have used that $\sum_i \dot{c}(\vec{x},t) = 0$ due to normalised concentrations. In the partial integration, from line 2 to 3, we have assumed no diffusion across system boundary, i.e., $\nabla c_i(\vec{x},t) = 0$ at the surface of the volume V.

Note that both terms in the integrand are negative. For the second term, this can be realised from the fact that if a component i has a concentration $c_i(\vec{x},t)$ larger than its equilibrium concentration c_{i0}, chemical reactions will lead to a decreasing concentration of i, reflected by a negative F_i, and the other way around for a compound with a lower concentration than equilibrium. This shows that the reaction-diffusion dynamics destroys the total information in two ways: Diffusion is active whenever $\nabla c_i(\vec{x},t) \neq 0$ and gives negative contribution to the change in information K. Chemical reactions are active whenever there is deviation from equilibrium,

$c_i(\vec{x}, t) \neq c_{i0}$ (for any i), and hence also contributes negatively to the change in information.

Let us discuss this briefly in a thermodynamic perspective. The chemical system, developing according to the reaction-diffusion dynamics, Eq. (7.21) without an in- and an out-flow (which will be treated separately), faces a change in a thermodynamic entropy, due to an entropy production σ (in units of Boltzmann's constant) according to

$$\sigma(\vec{x}, t) = \sum_i \left(D_i \frac{(\nabla c_i(\vec{x}, t))^2}{c_i(\vec{x}, t)} - \left(\ln \frac{c_i(\vec{x}, t)}{c_{i0}} \right) F_i(\vec{c}(\vec{x}, t)) \right) \geq 0 . \tag{7.26}$$

The thermodynamic entropy production is thus equal to the destruction of information derived above, with one part corresponding to the entropy produced due to diffusion within the system and one part given by the reactions that tend to even out the chemical non-equilibrium in the system. The entropy production certainly leads to a decay of the information in the system—decay of spatial information as well as of chemical information. This also illustrates that the exergy, being proportional to the total information, is destroyed by the rate of the thermodynamic entropy production.

7.2.5 Flows of Information in Closed Chemical Systems

We shall make an information-theoretic description of how information is flowing in the system that connects to the thermodynamic loss of information due to entropy production. This will be formulated in a continuity equation for information taking into account flows both in scale (r) and in space ($\vec{x}$), see Fig. 7.4.

In the limit of $r \to \infty$ we cannot distinguish any spatial structure, but the chemical information K_{chem} is still present, unaffected by the resolution parameter. In a decomposition of the total information, this part can therefore be considered as present at the limit $r \to \infty$. The chemical information will be consumed by the chemical reactions and we should therefore expect that a proper definition of information flow shows how information will flow in the direction towards smaller length scales r. If the system gives rise to spatial structure, that should be captured in the continuity equation, resulting in a temporal accumulation of information density $k(r, \vec{x}, t)$.

It is reasonable to think that information is leaving the system through the thermodynamic entropy production, at the finest length scales of the system, i.e., at $r = 0$. At this point information leaves the macroscopic description that we have of our system, and the information is spread out on microscopic degrees of freedom. This transfer of information was discussed in Sect. 7.1 as a transition phenomenon between the microscopic and the macroscopic descriptions. We thus define the information flow across length scales $j_r(r, \vec{x}, t)$ in the direction of smaller r, at the border $r = 0$, to be equal to the chemical entropy production,

$$j_r(0, \vec{x}, t) = \sigma(\vec{x}, t) \geq 0 . \tag{7.27}$$

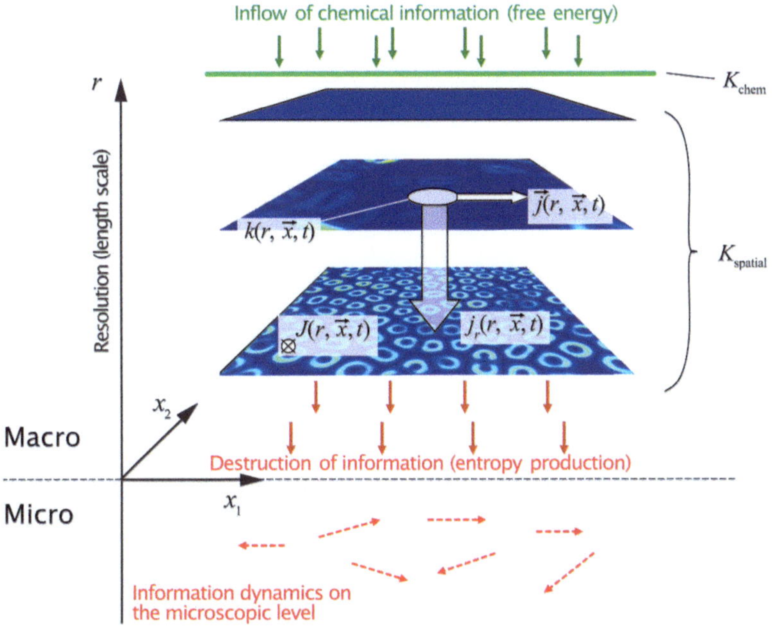

Fig. 7.4 A schematic picture of information (capacity) flows in a chemical pattern formation system. The pattern is characterised by an information density $k(r, \vec{x}, t)$ distributed over spatial dimensions as well as over different length scales r. Information is flowing both in space and in scale, where the flow is destroyed when it passes to the microscopic level. At this point, information disappears into microscopic degrees of freedom due to entropy production. In case of an open system, information capacity may enter the system at the largest length scales due to a diffusion-controlled inflow of chemical information or Gibbs free energy. A pattern is formed when information that flows downwards in scale is aggregated at certain positions as described by the continuity equation. Picture adapted from Lindgren (2015)

To define this flow for general resolution values r, we generalise by introducing the resolution operator into the expression for entropy production, and we define j_r by

$$j_r(r, \vec{x}, t) = \sum_i \left(D_i \frac{(\nabla \tilde{c}_i(\vec{x}, t))^2}{\tilde{c}_i(\vec{x}, t)} - \left(\ln \frac{\tilde{c}_i(\vec{x}, t)}{c_{i0}} \right) \tilde{F}_i(\vec{c}(\vec{x}, t)) \right) . \tag{7.28}$$

At infinitely bad resolution ($r \to \infty$) the reaction-diffusion dynamics, Eq. (7.24), captures the changes in average concentration

$$\dot{\bar{c}}_i(t) = \overline{F}_i(\vec{c}(\vec{x}, t) . \tag{7.29}$$

Using this, the information flow j_r, in the limit $r \to \infty$, can be written

$$
\begin{aligned}
j_r(\infty, \vec{x}, t) &= \sum_i \left(\left(\ln \frac{\overline{c}_i(t)}{c_{i0}} \right) \overline{F}_i(\vec{c}(\vec{x}, t)) \right) = \\
&= -\sum_i \dot{\overline{c}}_i(t) \ln \frac{\overline{c}_i(t)}{c_{i0}} \, . \qquad (7.30)
\end{aligned}
$$

Note that this expression is equal to the decay of chemical information, $-\mathrm{d}k_{\mathrm{chem}}/\mathrm{d}t$, since

$$
\begin{aligned}
\dot{k}_{\mathrm{chem}}(t) &= \frac{\mathrm{d}}{\mathrm{d}t} \sum_i \overline{c}_i \ln \frac{\overline{c}_i}{c_{i0}} = \sum_i \dot{\overline{c}}_i(t) \left(\ln \frac{\overline{c}_i(t)}{c_{i0}} + 1 \right) = \\
&= \sum_i \dot{\overline{c}}_i(t) \ln \frac{\overline{c}_i(t)}{c_{i0}} \, . \qquad (7.31)
\end{aligned}
$$

Here we have used $\sum_i \mathrm{d}c_i/\mathrm{d}t = 0$ due to the normalisation. In general it holds that $\mathrm{d}k_{\mathrm{chem}}/\mathrm{d}t \leq 0$ (except for extreme cases), which means that in general there is a positive flow of information $j_r(\infty, \vec{x}, t)$ at the infinite length scale limit in the direction of smaller length scales, similar to what we have at the other limit, $r = 0$. The interpretation is that information flows from its origin as chemical information down through the length scales, temporarily accumulating as spatial patterns are formed, eventually continuing to the finest length scales where it disappears as entropy production.

7.2.6 A Continuity Equation for Information in the Case of a Closed System

We can view the information density k as a generalised form of exergy (or free energy). By the flow across length scales j_r we have accounted for the destruction of information from entropy production in chemical reactions and diffusion within the system. We will now derive a continuity equation for information which will connect the previously defined flow across length scales, $j_r(r, \vec{x}, t)$, with the change of local information $k(r, \vec{x}, t)$ and a flow of information in space $\vec{j}(r, \vec{x}, t)$. For a closed chemical system such a continuity equation takes the form

$$
\dot{k}(r, \vec{x}, t) = r \frac{\partial}{\partial r} j_r(r, \vec{x}, t) - \nabla \cdot \vec{j}(r, \vec{x}, t) \, . \qquad (7.32)
$$

The equation states that, for a closed system, the local information density k changes due to accumulation of the information flows across length scales (in the downward direction) and across space. After several steps of derivation (which is left as a difficult exercise), this continuity equation then implies a definition of the $\nabla \cdot \vec{j}$ term,

$$\nabla \cdot \vec{j}(r, \vec{x}, t) = r\frac{\partial}{\partial r} j_r(r, \vec{x}, t) - \dot{k}(r, \vec{x}, t) = \cdots =$$
$$= -r^2\nabla^2 \sum_i \left(\tilde{F}(\vec{\tilde{c}}(\vec{x}, t)) \ln \frac{\tilde{c}_i(r, \vec{x}, t)}{c_{i0}} \right) . \tag{7.33}$$

If we require that $\vec{j}(r, \vec{x}, t) = 0$ when $\tilde{c}_i$ is uniform and that $\vec{j}$ is rotation-free (i.e., does not contain any term $\nabla \times \vec{A}$), the spatial flow $\vec{j}$ is defined

$$\vec{j}(r, \vec{x}, t) = -r^2\nabla \sum_i \left(\tilde{F}(\vec{\tilde{c}}(\vec{x}, t)) \ln \frac{\tilde{c}_i(r, \vec{x}, t)}{c_{i0}} \right) . \tag{7.34}$$

Note that the spatial flow depends on the presence of chemical reactions. This means that when reactions are not present, the only flow is the one across length scales. This is a direct consequence of the fact that the resolution operator, the Gaussian blur, is equivalent to a diffusion process. The flow across length scales, though, depends on both reactions and diffusion. This should be expected since it is a generalisation of the entropy production, and entropy is produced in both these processes.

7.2.7 A Continuity Equation for Information in the Case of an Open System

Finally, we open the system for inflow and outflow of chemical species, given by the term B in the reaction-diffusion dynamics, Eq. (7.22). In addition to the terms in the continuity equation for the closed system, already defined, we need to introduce a local source (or sink) term $J(r, \vec{x}, t)$ which will capture possible local effects from the system being open. A schematic picture is illustrating this in Fig. 7.4. The general continuity equation for an open system then takes the following form,

$$\dot{k}(r, \vec{x}, t) = r\frac{\partial}{\partial r} j_r(r, \vec{x}, t) - \nabla \cdot \vec{j}(r, \vec{x}, t) + J(r, \vec{x}, t) . \tag{7.35}$$

For an open system, there are two ways in which the information characteristics of the system is directly affected by the flow of chemical components across the system border. First, an inflow and an outflow of components affect the average concentrations of the transported components and in that way the chemical information is changed. For a driven chemical system, with an inflow of a fuel component and an outflow of a (waste) product, the flow across the system boundary typically keeps the chemical information at a level sufficient for driving the information flows in the system. But, there is also a direct influence the inflow may have on spatial patterns in the system. For example, diffusion into or out of the system of a component that has a spatial variation, directly leads to that spatial pattern being less accentuated, and in that way the flow destroys the local information density $k(r, \vec{x}, t)$.

The negative effect on information density k from the diffusion over the system boundary is captured by the sink term J in the continuity equation. For a diffusion controlled flow, as in Eq. (7.22), we get (after some calculations) the following expression for J,

$$J(r,\vec{x},t) = -\sum_i b_i \left(\tilde{c}_i(r,\vec{x},t) + c_i^{(\mathrm{res})}\right) (r\nabla \ln \tilde{c}_i(r,\vec{x},t))^2 \leq 0\,, \tag{7.36}$$

which shows that J is a sink for information. In conclusion, we have, see Eqs. (7.18), (7.28), (7.34), and (7.36),

$$k(r,\vec{x},t) = r^2 \sum_i \frac{(\nabla \tilde{c}_i(r,\vec{x}))^2}{\tilde{c}_i(r,\vec{x})}\,, \tag{7.37}$$

$$j_r(r,\vec{x},t) = \sum_i \left(D_i \frac{(\nabla \tilde{c}_i(\vec{x},t))^2}{\tilde{c}_i(\vec{x},t)} - \left(\ln \frac{\tilde{c}_i(\vec{x},t)}{c_{i0}}\right) \tilde{F}_i(\vec{c}(\vec{x},t)) \right)\,, \tag{7.38}$$

$$\vec{j}(r,\vec{x},t) = -r^2 \nabla \sum_i \left(\tilde{F}(\vec{c}(\vec{x},t)) \ln \frac{\tilde{c}_i(r,\vec{x},t)}{c_{i0}} \right)\,, \tag{7.39}$$

$$J(r,\vec{x},t) = -\sum_i b_i \left(\tilde{c}_i(r,\vec{x},t) + c_i^{(\mathrm{res})}\right) (r\nabla \ln \tilde{c}_i(r,\vec{x},t))^2\,. \tag{7.40}$$

These quantities can be integrated over either position space $\vec{x}$ or resolution length scale r in order to derive aggregated information quantities like $k(r,t)$, i.e., the resolution-dependent information, or $k(\vec{x},t)$ and $\vec{j}(\vec{x},t)$ which are the normal information density and the corresponding information flow. If we integrate the continuity equation, Eq. (7.35), over resolution lengths r, we get the following balance equation for the information density $k(\vec{x},t)$,

$$\dot{k}(\vec{x},t) + \nabla \cdot \vec{j}(\vec{x},t) - J(x,t) + \sigma(\vec{x},t) + \dot{k}_{\mathrm{chem}}(\vec{x},t) = 0\,. \tag{7.41}$$

Here the entropy production and the change in chemical information come from the upper and lower limits of the flow j_r.

In the next section these information-theoretic concepts will be applied to a simple model of a chemical self-organizing system.

7.3 Application to the Self-replicating Spots Dynamics

We apply the formalism to the pattern formation of the "self-replicating spots" system (Gray and Scott 1984; Lee et al. 1993; Pearson 1993),

$$\mathrm{U} + 2\mathrm{V} \rightleftarrows 3\mathrm{V}\,. \tag{7.42}$$

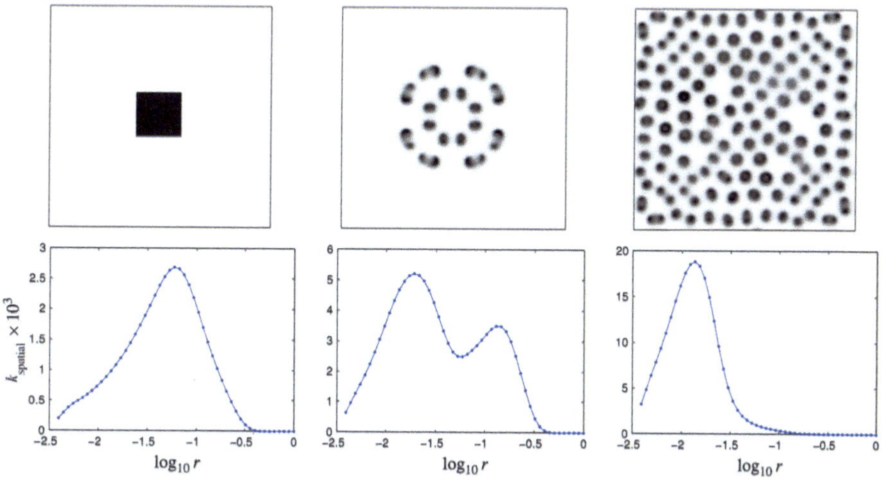

Fig. 7.5 Top row: The concentration of the chemical V in the system at three times; $t = 0, t = 1000$, and $t = 7000$ steps. White corresponds to zero concentration, and black corresponds to a concentration of one half. Bottom row: The spatial information $k_{\text{spatial}}(r)$, as a function of the resolution r, for the same time steps as above. The length of the system is 1, $D_U = 2$, $D_V = 0.05$, $\hat{D} = 0.02$, $k = 0.058$. Figure from Lindgren et al. (2004)

Let $c_U = c_U(\vec{x}, t)$ and $c_V = c_V(\vec{x}, t)$ denote the spatially dependent concentrations of U and V, respectively. The dynamics of the concentrations is then given by reaction-diffusion equations, cf. Eq. (7.21),

$$\dot{c}_U = D_U \nabla^2 c_U - (c_U - k_{\text{back}} c_V) c_V^2 + \hat{D}(1 - c_U) , \quad (7.43)$$

$$\dot{c}_V = D_V \nabla^2 c_V + (c_U - k_{\text{back}} c_V) c_V^2 - (\hat{D} + k) c_V , \quad (7.44)$$

where $\hat{D}$ and k are constants characterising the diffusion based exchange with the reservoir. We have introduced a very slow back reaction ($k_{\text{back}} = 10^{-5}$) in order to get the relationship between equilibrium concentrations of U and V defined by the reactions.

In Fig. 7.5, the dynamics is illustrated starting from an initial state (left) with a square of high concentration of V. As the system evolves four concentration peaks (spots) emerge from the corners of the square, and these spots reproduce by growing and splitting until the system is filled with spots (middle and right). In the process, spots may disappear, which leaves space for other spots to reproduce. In the lower part of the figure, the decomposition of the information in the pattern with respect to scale is plotted for the three snapshots above (at time 0, 1000, and 7000, respectively).

It is clear that the initial state has a longer characteristic length scale as detected by the information density. When the system produces the spots at the significantly shorter length scale, information is found both at the old length, now due to the size of the cluster (middle), and at the length scale of the spots. When the square

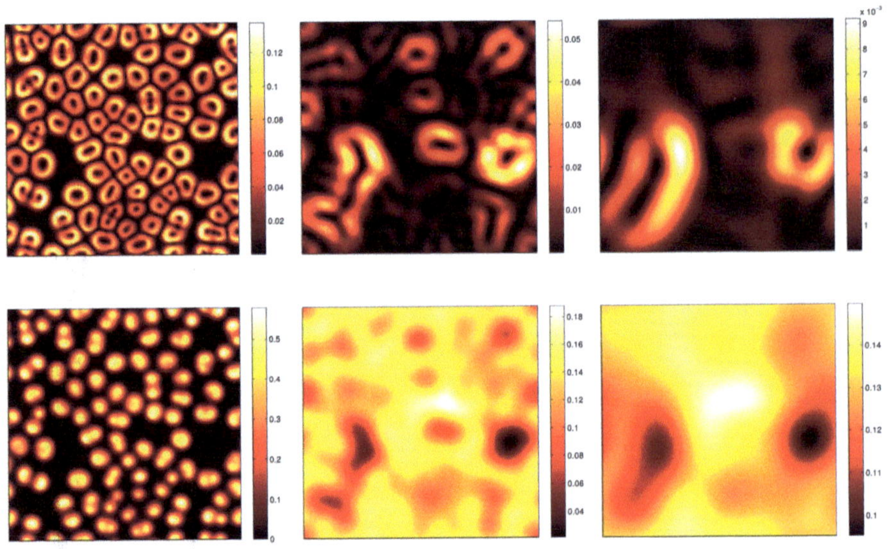

Fig. 7.6 Top row: The information density $k(r, \vec{x}, t)$ in the system at $t = 10000$ steps, for three values of the resolution, $r = 0.01$, $r = 0.05$, and $r = 0.10$. Bottom row: The corresponding information flow $j_r(r, \vec{x}, t)$. The length of the system is 1, $D_U = 2$, $D_U = 0.05$, $\hat{D} = 0.02$, $k = 0.058$. Figure from Lindgren et al. (2004)

distribution has been completely decomposed into spots, no information is left at the initial length scale.

In Fig. 7.6, we show the information density over the system for three different length scales after long time (upper part), and the information flow in scale, j_r, for the same state (lower part). At low resolution, or large r (right), the information density is low and captures structures of longer lengths, while at finer resolution, small r (left), the information density is large and reflects the pattern of spots. Note that each spot is seen as a circle in the information density picture, since the information density is sensitive to gradients in the pattern.

At worse resolution levels, e.g., $r = 0.05$, the information density does not capture the individual spots. Instead, the empty areas, where spots have disappeared, contribute with a higher information density as these are characterised by larger length scales.

The information flow in scale, j_r, is close to homogenous for large r (right), but when information moves on to finer scales of resolution, the spatial flow $\vec{j}$ leads to a modification of the information flow so that a higher flow j_r is obtained at the concentration peaks. At finest resolution, $r = 0$, the information leaves the system as entropy production, which is mainly located to the concentration peaks where the chemical activity is high.

Chapter 8
Chaos and Information

Abstract Chaotic systems are characterised by their sensitivity to small disturbances in initial position, which results in diverging trajectories, with the exponential rate of divergence given by the *Lyapunov exponent*. This amplification of the details in the initial position can be seen as a flow of information from smaller to larger length scales in state space, and it can be quantified using information theory. This flow can also be quantified as an entropy that is generated by the chaotic system, despite the fact that the dynamics is deterministic. In order to make an information-theoretic analysis of chaotic dynamical systems, we describe the system trajectory as a symbol sequence, which is referred to as *symbolic dynamics*. If the description of the chaotic dynamics in terms of symbol sequences is properly done, the full characteristics of the dynamical system can be derived from the symbol sequence description. In chemical self-organizing systems one can identify a tendency of information to flow from larger to smaller length scales, as we have discussed in the previous chapter. Still, it is clear that noise or fluctuations are important to break symmetries and to initiate the formation of spatial structure. Quantitatively this information flow from the noise, or from the microscales of the system, is very small and is hidden by the much larger thermodynamically related flow towards the microscales that is present in chemical dynamics. In chaotic systems, noise is of an even higher importance. Chaos may even be characterised by the extent to which a system is sensitive to noise. The Lyapunov exponents of a dynamical system quantifies how noise is amplified in the dynamics. In this chapter, we shall make an information-theoretic interpretation of these exponents and relate them to an entropy concept in dynamical systems, *the measure entropy*.

8.1 Basic Dynamical Systems Concepts

We start with a short condensed introduction to some basic concepts in discrete time dynamical systems—sufficient for making the connection to the information-theoretic concepts we have introduced before, especially to the ones presented in Chap. 3 on symbol sequences. For a more thorough introduction, there are several text books one may consult, for example (Strogatz 2018).

© Springer-Verlag GmbH Germany 2024

K. Lindgren, *Information Theory for Complex Systems*, Understanding Complex Systems, https://doi.org/10.1007/978-3-662-68214-2_8

8.1.1 Iterated Maps, Fixed Points, and Periodic Orbits

In this presentation, we only consider time-discrete systems in one variable, where the state $x(t) \in R$, with a dynamics given by the an *iterated map*

$$x(t+1) = f(x(t)) \ . \tag{8.1}$$

The function f that determines the next time step is a differentiable real-valued function on R, $f : R \to R$. The formalism can be extended to higher dimensions and to continuous time, see for example the classic review by Eckmann and Ruelle (1985).

A dynamical system of this type has a number of possible types of behaviour. The system can be in a fixed point, $x^* = f(x^*)$, and nothing changes. The trajectory of the system may also be on a cycle with a certain period T, characterised by $x(T) = f^T(x(0)) = x(0)$ and $x(k) \neq x(0)$ for $0 < k < T$. Fixed points and periodic orbits may be stable or unstable. Instability means that an arbitrarily small disturbance is sufficient for taking the system away from the point (or the orbit). A fixed point x^* is unstable if $|f'(x^*)| > 1$, since any disturbance will result in a trajectory that moves away from x^*. Similarly, a periodic orbit of period T is unstable if the corresponding fixed point of f^T is unstable.

The dynamics can be said to be chaotic if the trajectory does not approach a fixed point or a periodic orbit and if it does not diverge. In this case, the trajectory of the system is approaching a (usually complex) set of points in state space, called a *strange attractor*.

8.1.2 Probability Densities and Measures on the State Space

Dynamical systems are often characterised by quantities that are averages over either time or state space (here R). Using the terminology and notation from dynamical systems, such a spatial average is based on a probability measure μ on the state space, such that $\mu(E)$ can be interpreted as the probability for finding the system in a certain subset $E \subseteq R$. We are usually interested in an *invariant measure*,[1] i.e., a probability measure that does not change under the dynamics of Eq. (8.1),

$$\mu(f^{-1}(E)) = \mu(E) \ , \tag{8.2}$$

where E is a set of points in R, and $f^{-1}(E)$ is the set of points that under the map f is transformed to E, i.e., $f^{-1}(E) = \{x : f(x) \in E\}$. This can also be expressed as $f(f^{-1}(E)) = E$. There may be several invariant measures for a dynamical system. If there is a fixed point x^*, then a point distribution $\delta(x - x^*)$ in that point is an

[1] This corresponds to a stationary probability distribution when one has a system with a finite state space, for example a finite state automaton.

invariant measure, even if the fixed point is unstable. Any linear combination of invariant measures is also an invariant measure.

In order to eliminate invariant measures that correspond to unstable modes of the dynamics one may introduce the so-called *physical measure*, by adding noise to the dynamics and using the resulting measure when reducing the noise to zero. In this way, the physical measure does not include contributions from any of the unstable modes (unstable fixed points or unstable periodic orbits), but if the system is chaotic the resulting measure is reflecting how the states are spread over the attractor. The physical measure can be composed of several invariant measures (for example from several stable fixed points or periodic orbits).

If an invariant measure cannot be decomposed into parts that are invariant, the measure is called *ergodic*. The implications of this is that an average (of any function ϕ) calculated using the ergodic measure is identical (almost always) to a temporal average following the trajectory of the system, $(f(x(0)), f^2(x(0)), f^3(x(0)), \ldots)$. This is the ergodicity theorem,

$$\int \mathrm{d}x\, \mu(x)\phi(x) = \lim_{T\to\infty} \frac{1}{T} \sum_{k=1}^{T} \phi(f^k(x(0)))\;, \tag{8.3}$$

for almost all initial states $x(0)$. This is analogous to the connection between a probabilistic and a trajectory based average that was discussed for the ergodicity property in stochastic processes in Chap. 3, see Eq. (3.7).

8.1.3 Lyapunov Exponent

The trajectory in a chaotic system is sensitive to small disturbances in the initial state. The tendency to amplify small perturbations is quantified by the *Lyapunov exponent* of the system (or several exponents in the case of systems in higher dimensions).

Suppose that there is a small change $\delta x(0)$ in the initial state $x(0)$. By using a linear expansion, we find that at time t this has led to a change $\delta x(t)$ compared to the point at time t starting from $x(0)$, given by

$$\begin{aligned}\delta x(t) &\approx \delta x(0) \left| \frac{\mathrm{d}f^t}{\mathrm{d}x}(x(0)) \right| = \\ &= \delta x(0) \left| f'(x(t-1))\, f'(x(t-2)) \ldots f'(x(0)) \right| \;,\end{aligned} \tag{8.4}$$

where we have used the chain rule to expand the derivative of f^t. In the limit of infinitesimal perturbations $\delta x(0)$ and infinite time, we get an average exponential amplification, the Lyapunov exponent λ,

$$\lambda = \lim_{t\to\infty} \frac{1}{t} \ln \left| \frac{\delta x(t)}{\delta x(0)} \right| = \lim_{t\to\infty} \frac{1}{t} \ln \left| \frac{\mathrm{d} f^t}{\mathrm{d} x}(x(0)) \right| =$$

$$= \lim_{t\to\infty} \frac{1}{t} \sum_{k=0}^{t-1} \ln \left| f'(x(k)) \right| . \tag{8.5}$$

If the system is characterised by an ergodic measure μ we can use the ergodicity theorem to express the Lyapunov exponent as

$$\lambda = \lim_{t\to\infty} \frac{1}{t} \sum_{k=0}^{t-1} \ln \left| f'(x(k)) \right| = \int \mathrm{d}x \, \mu(x) \ln \left| f'(x) \right| . \tag{8.6}$$

If the Lyapunov exponent is larger than zero, the system amplifies small perturbations and we have chaos, while if the exponent is negative the system is stable, characterised by a fixed point or a periodic orbit.

8.1.4 The Lyapunov Exponent as an Information Flow from "Micro" to "Macro"

An illustration of the Lyapunov exponent as an information flow from smaller to larger length scales in state space is given by the following example. Suppose that we determine the position of the system at a certain time t_0, with some resolution δ, implying that we can describe the position as a uniform probability density over a certain interval $I_0 = [x_0 - \delta, x_0 + \delta]$. In n time steps, the dynamics will roughly transform this uncertainty interval to

$$I_n \approx \left[f^n(x_0) - \delta \left| \frac{\mathrm{d} f^n}{\mathrm{d} x}(x_0) \right| , \; f^n(x_0) + \delta \left| \frac{\mathrm{d} f^n}{\mathrm{d} x}(x_0) \right| \right] , \tag{8.7}$$

if δ is small enough. Assuming a chaotic system, the new interval will be larger, which means that we have a less good descriptions of the future position of the system, based on our information of the position at time t_0 and the knowledge about the dynamics, Eq. (8.1). Let us use this as an a priori description of the system position at this future point in time, using a uniform probability distribution $p_0(x) = p_0$ in the interval I_n that (by normalisation) equals

$$p_0 = \frac{1}{2\delta \left| \frac{\mathrm{d} f^n}{\mathrm{d} x}(x_0) \right|} . \tag{8.8}$$

In these n time steps the dynamics will bring the system to some point within this interval. If we observe the system at the new time, using the same level of resolution, we get a new (better) estimate of the position, characterised by a smaller interval (length 2δ) and a uniform probability $p = 1/(2\delta)$. The relative information associated with this observation is then

$$K[p_0; p] = \int \mathrm{d}x\ p \ln \frac{p}{p_0} = \ln \left| \frac{\mathrm{d}f^n}{\mathrm{d}x}(x_0) \right| . \tag{8.9}$$

In the limit of arbitrarily fine resolution and infinite time, the information I gained per time step is

$$I = \lim_{n\to\infty} \frac{1}{n} K[p_0; p] = \lim_{n\to\infty} \ln \left| \frac{\mathrm{d}f^n}{\mathrm{d}x}(x_0) \right| = \lambda . \tag{8.10}$$

This illustrates the interpretation of the Lyapunov exponent λ as a measure of information flow I from smaller to larger length scales in state space. This also means that the observation of the state at time t gives information about details of the starting point at time 0, not initially detectable at the given resolution.

8.2 Dynamical Systems Entropy and Information Flow

In earlier chapters we have characterised disorder in symbol sequences by the use of entropies. We shall make a similar approach here, to characterise the noise amplification in dynamical systems in information-theoretic terms. This presentation builds on the formalism presented in the review by Eckmann and Ruelle (1985).

Consider again a dynamical system defined by the iterated map $f(x)$, and suppose that it generates (or is characterised by) an ergodic measure μ. In order to transform the dynamics in the continuous state space into a dynamical system of discrete symbols, we make a decomposition, or a *partition*, of the state space into r non-overlapping subsets. The partition is denoted $\vec{A} = (A_1, A_2, ..., A_r)$ and it covers the whole state space. A trajectory in state space then jumps between these subsets which can be expressed as a sequence of discrete states (or subsets).

For each subset A_j we let $f^{-k}(A_j)$ denote the set of points that after k steps end up in A_j, i.e., $f^k(f^{-k}(A_j)) = A_j$. Based on this we construct a new decomposition $B^{(n)}$ of the state space, using the subsets $B_{i_1 \ldots i_n}$, defined by

$$B_{i_1 \ldots i_n} = A_{i_1} \cap f^{-1}(A_{i_2}) \cap \ldots \cap f^{-n+2}(A_{i_{n-1}}) \cap f^{-n+1}(A_{i_n}) , \tag{8.11}$$

where the indices $i_k \in \{1, ..., r\}$. A subset B_σ in $B^{(n)}$ is then characterised by a sequence of indices $\sigma = i_1, ..., i_n$. The interpretation is as follows. If the system starts at time $t = 1$ with a state in B_σ, the system will in n consecutive time steps be found in the subsets: A_{i_1} at time $t = 1$, A_{i_2} at time $t = 2$, etc., until it finally reaches A_{i_n} at time $t = n$, i.e., $x(t) \in A_{i_t}$. This means that the sequence $\sigma = i_1, ..., i_n$ can be viewed as a symbol sequence generated by the dynamics, given a certain basic partition $\vec{A}$ of the state space.

In other words, the components in the partition $B^{(n)}$ are the subsets B_σ. Each subset corresponds to certain n-length symbol sequences $\sigma = i_1, \ldots, i_n$ generated by the dynamics. The ergodic measure μ of the dynamics then implies that each subset B_σ is characterised by a certain probability, $\mu(B_\sigma)$, and we use this to define an entropy H for the partition $B^{(n)}$,

$$H(B^{(n)}) = \sum_\sigma \mu(B_\sigma) \ln \frac{1}{\mu(B_\sigma)} \,. \tag{8.12}$$

This corresponds to the entropy for symbol sequences of length n, the block entropy, Eq. (3.8). We now define an entropy per time step (or symbol), $h(\mu, \vec{A})$, in the same way as we defined the entropy of a stochastic process in Eq. (3.9), by taking the limit of infinite time (or symbol block length),

$$h(\mu, \vec{A}) = \lim_{n\to\infty} \frac{1}{n} H(B^{(n)}) = \lim_{n\to\infty} \left(H(B^{(n+1)}) - H(B^{(n)}) \right) \,. \tag{8.13}$$

This entropy depends on how we have chosen to make the partition $\vec{A}$, i.e., what points in state space the symbols represent, and which invariant measure μ that is used. Next, we take the limit of this entropy when the partition is made arbitrarily fine, i.e., the size of the subsets in $\vec{A}$ tends to zero, which leads to the definition of the *measure entropy*,

$$s_\mu = \lim_{\mathrm{diam}(\vec{A})\to 0} h(\mu, \vec{A}) \,, \tag{8.14}$$

where $\mathrm{diam}(\vec{A})$ denotes the largest distance between two points in any subset $A_j \in \vec{A}$. Note that in this limit we increase the number of symbols in the alphabet to infinity.

The entropy $h(\mu, \vec{A})$ is the average entropy per time step when observing the symbols (defined by the partition $\vec{A}$) that are generated by the dynamics. If the system stabilises at a fixed point or a periodic orbit the symbol sequence will certainly have zero entropy. If the system is chaotic, the result may depend on how the partition is made, and therefore the limit used in the definition of measure entropy is needed. If $s_\mu > 0$, we cannot (always) tell which part of $\vec{A}$ the system will visit in the next time step, regardless of how many previous steps (symbols) we have observed, and regardless of how fine we make the partition. Thus, there is a "creation" of information through the dynamics, and s_μ can be said to quantify the average rate of creation of information.

In some cases the limit $\mathrm{diam}(\vec{A}) \to 0$ is not necessary, but a finite partition may be sufficient. If the partition $\vec{A}$ is done so that $\mathrm{diam}(B^{(n)}) \to 0$ when $n \to \infty$, then the measure entropy is determined already by the entropy based on $\vec{A}$, Eq. (8.13). If this holds, $\vec{A}$ is a *generating partition*. In this case we have a finite decomposition of state space, each part represented by a certain symbol. This is sufficient for providing a full characterisation of the dynamics. The dynamics $x(t)$ is then represented by a sequence of symbols $s(t)$. The longer sequence we observe, for a chaotic system, the more accurate will our knowledge be about the real position in state space of the system at time $t = 0$. Each time step may bring us new information, in average the information quantity s_μ. A chaotic system has $s_\mu > 0$, and this can be viewed as an information flow from smaller to larger length scales in state space. It is reasonable to expect that if the Lyapunov exponent corresponds to a divergence of a factor of two ($\lambda = \ln 2$), then there is one bit of information creation per time step ($s_\mu = \ln 2$). In fact, it often holds (in all examples that we will discuss) that the measure entropy equals the Lyapunov exponent λ if it is larger than 0,

$$s_\mu = \lambda \,. \tag{8.15}$$

If $\lambda < 0$, though, the measure entropy is zero, $s_\mu = 0$, since the symbol sequence that is produced will be periodic (or consists of the repetition of one symbol). Equation (8.15), is an instance of *Pesin's theorem* stating that the measure entropy equals the sum of the positive Lyapunov exponents, valid also in higher dimensions.

8.2.1 Extended Example of a Generating Partition for a Skew Roof Map

Consider an iterated map defined by $f(0) = a$, $f(a) = 1$, $f(1) = 0$ (where $0 < a < 1$), with $f(x)$ being linear in the intervals $[0, a]$ and $[a, 1]$, as illustrated in Fig. 8.1. By defining a partition so that points x in the first interval generate symbol A and in the second interval symbol B, the iterated map is transformed into a symbolic dynamics with the alphabet {A, B}. We let the symbols denote the intervals they correspond to: $\mathrm{A} = [0, a]$ and $\mathrm{B} =]a, 1]$. Following the procedure summarised in Eq. (8.11), we construct the subsets of A and B corresponding to increasing lengths of sequences generated in the symbolic dynamics. For example, the set BBA is the subset that starts in B, in one iteration maps to B again, and in the next iteration is mapped to A,

$$\mathrm{BBA} = \mathrm{B} \cap f^{-1}(\mathrm{B}) \cap f^{-2}(\mathrm{A}) \,. \tag{8.16}$$

By taking such sequences $x_1 x_2 ... x_m$ of increasing length m, where $x_j \in \{\mathrm{A}, \mathrm{B}\}$, we see that the corresponding sets decreases in size, formally expressed by $\mathrm{diam}(x_1 x_2 ... x_m) \to 0$, as $m \to \infty$. This means that we have a generating partition.

In this system, with the present choice of partition, we have a have a simple way in which the sets A and B map onto these sets or the union of them. We note that the set A maps onto the set B, so that if we have a uniform probability density (uniform measure) over A, that is evenly distributed in one iteration over the set B. Similarly, a uniform measure over B is evenly distributed over A∪B, the whole unit interval, in one iteration. The fraction of the probability density from B that ends up in A is then $|\mathrm{A}|$, i.e., the length of the A interval, and the corresponding fraction for B is $|\mathrm{B}|$. Because of this, we can easily find an invariant distribution, since the requirement on invariant measure from Eq. (8.2),

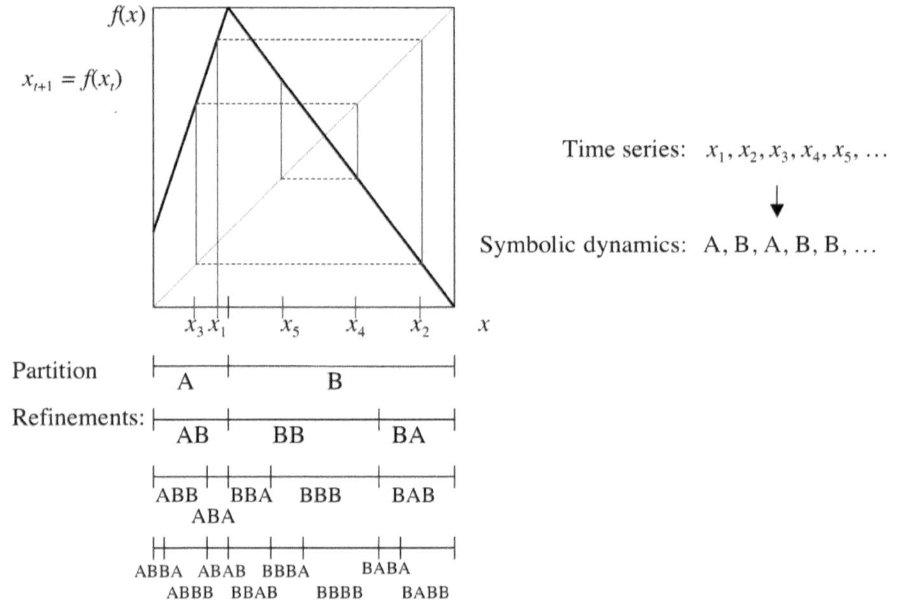

Fig. 8.1 The graph shows an iterated map with $f(0) = a$, $f(a) = 1$, $f(1) = 0$, along with a partition associating points to the left and to the right of the peak with A and B, respectively. Two the right the corresponding symbolic dynamics is illustrated. The sets corresponding to A and B are then refined by finding subsets that in one iteration are mapped to the A and B, respectively. In this case it turns out that no points in A are mapped to A so there is no such subset. A continued refinement leads the sizes of subsets decreasing towards 0, and we have a generating partition

$$\mu\left(f^{-1}(\mathrm{B})\right) = \mu(\mathrm{B}) \ , \tag{8.17}$$

$$\mu\left(f^{-1}(\mathrm{A})\right) = \mu(\mathrm{A}) \ , \tag{8.18}$$

together with the result for this specific map f

$$\mu\left(f^{-1}(\mathrm{B})\right) = \mu(\mathrm{A}) + |\mathrm{B}|\mu(\mathrm{B}) \ , \tag{8.19}$$

$$\mu\left(f^{-1}(\mathrm{A})\right) = |\mathrm{A}|\mu(\mathrm{B}) \ , \tag{8.20}$$

results in the equations

$$\mu(\mathrm{B}) = \mu(\mathrm{A}) + |\mathrm{B}|\mu(\mathrm{B}) \ , \tag{8.21}$$

$$\mu(\mathrm{A}) = |\mathrm{A}|\mu(\mathrm{B}) \ , \tag{8.22}$$

The constraint that the total measure is 1, $\mu(\mathrm{A}) + \mu(\mathrm{B}) = 1$, replaces the last equation and we find that $\mu(\mathrm{A}) = (1 - |\mathrm{B}|)/(2 - |\mathrm{B}|)$ and $\mu(\mathrm{B}) = 1/(2 - |\mathrm{B}|)$. This procedure, with its solution, is equivalent to our analysis of finite state automata and the stationary distribution over the nodes, discussed in Chap. 3. In this example, the sym-

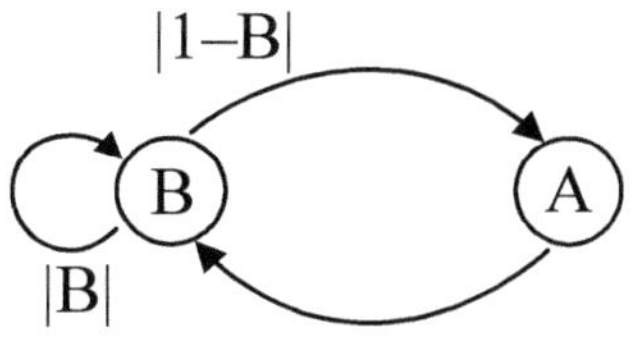

Fig. 8.2 The finite state automaton representing the symbolic dynamics of Fig. 8.1 with an invariant measure that is uniform within A and B

bolic dynamics is in the form of a Markov process, defined by Eqs. (8.21) and (8.22), and illustrated in Fig. 8.2. This type of partition, resulting in a Markov process for the symbolic dynamics is called a *Markov partition*. The invariant measure for the sets A and B corresponds to the stationary probability distribution over the nodes A and B, respectively.

This means that the measure entropy s_μ, for this invariant measure, equals the entropy s of the stochastic process of Fig. 8.2. Based on our previous discussion on these processes, see Eq. (3.22), we conclude that

$$s_\mu = s = \mu(B)\left(|B| \ln \frac{1}{|B|} + (1 - |B|) \ln \frac{1}{1-|B|}\right) . \tag{8.23}$$

From Eq. (8.15), we can determine the Lyapunov exponent, $\lambda = s_\mu$, since $s_\mu > 0$.

When we know the invariant measure, we can also calculate the Lyapunov exponent directly from Eq. (8.6), using the fact that the slope $|f'(x)|$ equals $(1 - |A|)/|A| = |B|/(1 - |B|)$ in A and $1/|B|$ in B. Then λ can be written

$$\begin{aligned}\lambda &= \int \mathrm{d}x\, \mu(x) \ln |f'(x)| = \int_A \mathrm{d}x\, \mu(x) \ln \frac{|B|}{1-|B|} + \int_B \mathrm{d}x\, \mu(x) \ln \frac{1}{|B|} = \\ &= \frac{1-|B|}{2-|B|} \ln \frac{|B|}{1-|B|} + \frac{1}{2-|B|} \ln \frac{1}{|B|} = s .\end{aligned} \tag{8.24}$$

To be formally correct, we need to note the following: The measure μ that we have used is actually not ergodic, so in principle it appears incorrect to use Eq. (8.6) to derive the Lyapunov exponent. The measure, being piece-wise uniform over the whole interval, includes points that correspond to the unstable fixed point and unstable periodic orbits, for example $x = 0$, $x = a$, and $x = 1$. And the measure defined for any such mode is in itself also invariant. So the overall invariant measure we have used can be decomposed into a number of invariant measures, and therefore it does not fulfil the requirement for ergodicity. But, since all these points of the periodic orbits, being part of the support of the measure μ, are of measure 0 (a countable set of points within an uncountable set), it still works to treat μ as if it is ergodic for calculation of the Lyapunov exponent.

Next, we discuss in what way a badly chosen partition leads to a symbolic dynamics that is not capturing the characteristics of the dynamical system.

8.2.2 An Example of a Partition that is not Generating

Consider the tent map, $f(x) = 1 - |2x - 1|$, see Fig. 8.3. Since this is a map with slope $|f'| = 2$ everywhere, we know that all fixed points and periodic orbits are unstable. Furthermore, Eq. (8.6) says that any ergodic invariant measure results in a Lyapunov exponent $\lambda = \ln 2$. Then, from Eq. (8.15) we also know that the measure entropy is the same, $s_\mu = \ln 2$.

All this can be verified with the approach above, for example, by making a partition dividing the unit interval into two equal halves. Here, though, we will discuss what happens when one chooses a different partition, that is not necessarily generating. Therefore, we make a partition by dividing the unit interval at $x = 3/4$, so that A is the interval $[0, 3/4]$ and B is $[3/4, 1]$. This partition with two steps of refinements are shown in Fig. 8.3. We note already at symbol sequences of length 2 that there is a set AA (corresponding to points for which the map generates a pair of A's in two iterations) that is not connected. A further refinement, going to symbol sequences of length 3 shows that this set is further split up in disconnected parts and same holds now for the set AAB. This means that there will exist sequences $x_1 x_2 ... x_m$ for which we do not have $\text{diam}(x_1 x_2 ... x_m) \to 0$, as $m \to \infty$. (The largest distance between two points in the set will typically be finitely separated since many sequences will have disconnected parts spread out over the unit interval.) Thus, this partition is not generating.

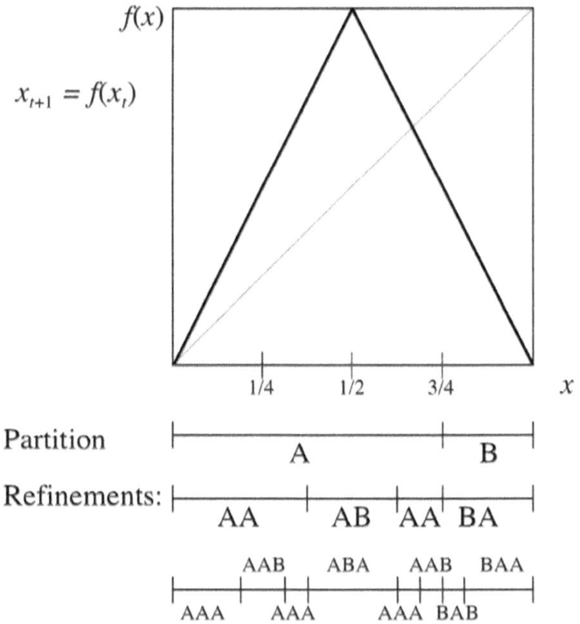

Fig. 8.3 The partition (A, B), applied to the tent map, is not generating

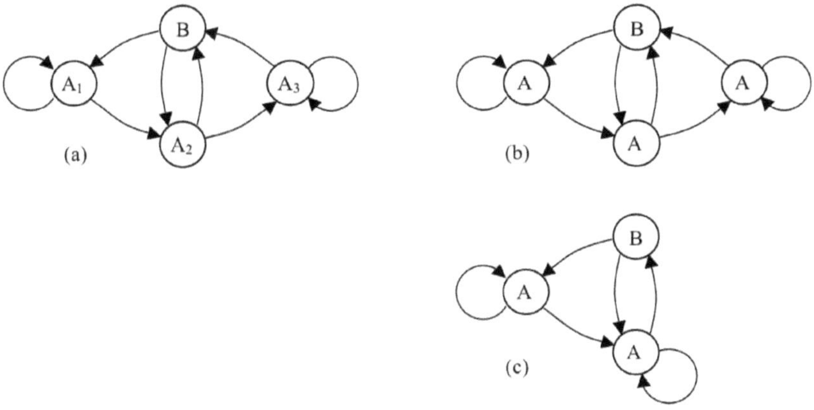

Fig. 8.4 **a** The finite state automaton corresponding to the symbolic dynamics of the generating partition (A_1, A_2, A_3, B). **b** The hidden Markov model corresponding to the symbolic dynamics of the non-generating partition (A, B) of Fig. (8.3), and **c** an equivalent representation of the hidden Markov model

The symbolic dynamics resulting from this partition can still be analysed using the approach of the previous section. If the interval A is divided in three equal parts (at the points $x = 1/4$ and $x = 1/2$), A_1, A_2, and A_3, then the partition (A_1, A_2, A_3, B) is in fact generating. From the tent map we see that these intervals, in one iteration, are mapped as follows,

$$A_1 \to A_1 \cup A_2 \ , \tag{8.25}$$

$$A_2 \to A_3 \cup B \ , \tag{8.26}$$

$$A_3 \to A_3 \cup B \ , \tag{8.27}$$

$$B \to A_1 \cup A_2 \ . \tag{8.28}$$

This means, using similar arguments as for the skew roof map, that the partition results in the symbolic dynamics being a Markov process with an automaton representation as in Fig. 8.4a. Since all intervals are of equal length, the transition probabilities are all $1/2$. This Markov process has an entropy of $\ln 2$, and gives the correct value for s_μ and thus also for λ.

The non-generating partition (A, B) has a symbolic dynamics that results from changing any symbol A_k to A. This means that the symbolic dynamics is a Hidden Markov model, shown in Fig. 8.4b, with an equivalent, more compact, representation in Fig. 8.4c. Even though the entropy involved in the choices in this automaton is still $\ln 2$, the entropy of the symbol sequence (the Hidden Markov model) is smaller. This follows from the fact that there are different combinations of choices in the two A nodes (in Fig. 8.4c) that result in the same symbol sequence, i.e., some of the choices made does not show up in the symbol sequence, and the entropy of this choice of symbolic dynamics is then smaller than $\lambda = \ln 2$.

Problems

8.1

Let a map $f(x)$ be defined by the figure below, where $f(1/3) = 1$, $f(2/3) = 2/3$, and $f(0) = f(1) = 0$.

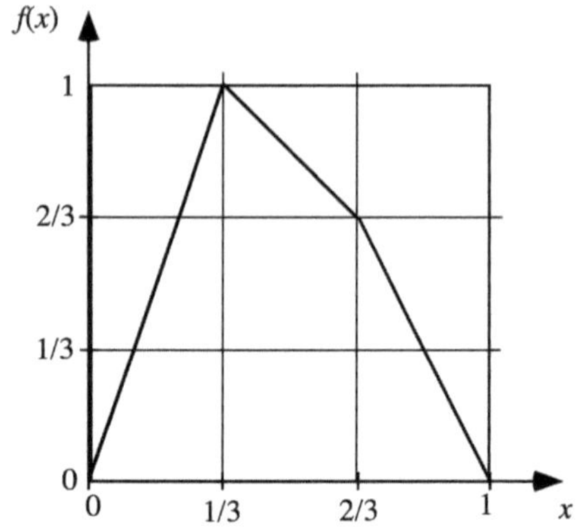

Consider the dynamical system $x_{t+1} = f(x_t)$.
(a) Find the invariant measure μ that characterises the chaotic behaviour, and determine the corresponding Lyapunov exponent λ by using that measure.
(b) Show that there is a partition that has a symbolic dynamics with a measure entropy s_μ that equals the Lyapunov exponent (as one should expect).
(c) Suppose that we at a certain time t observe the system in the region given by $x > 2/3$. If we find the system in this region again two time steps later, how much information do we gain by this observation?

8.2

Let a piecewise linear map $f(x)$ be defined by the figure below, where $0 < \beta < 1$, and where the map is determined by $f(0) = 3/4$, $f(3/4) = 1$, and $f(1) = \beta$.

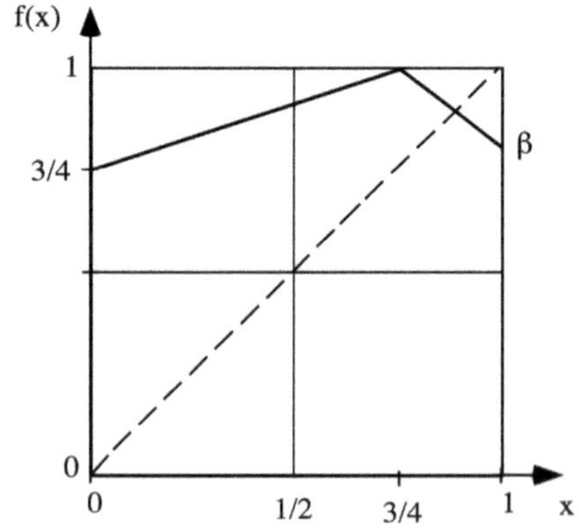

Consider the dynamical system $x_{t+1} = f(x_t)$.
(a) At what value on β does the system become chaotic (starting with β being close to 1)? Characterise the dynamics for β above the critical value for chaos (stable/unstable fixed point and/or periodic). Is there any other value on β that corresponds to a change in the dynamics in the non-chaotic regime?
(b) Suppose that $\beta = 0$. Determine the invariant measure that characterises the chaotic behaviour, and calculate the Lyapunov exponent λ. Calculate also the measure entropy from the finite state automaton describing the symbolic dynamics (for a generating partition).
(c) If you know that the system is in the region $x > 3/4$ at time t, how much information do you get if you observe the system in the region $x < 3/4$ at time $t + 3$ (again assuming $\beta = 0$)?

8.3
Let a piecewise linear map $f(x)$ be defined by the figure below, where $0 < \alpha < 1$, and where the map is determined by $f(0) = \alpha$, $f(1/4) = 0$, and $f(1) = 1/4$.

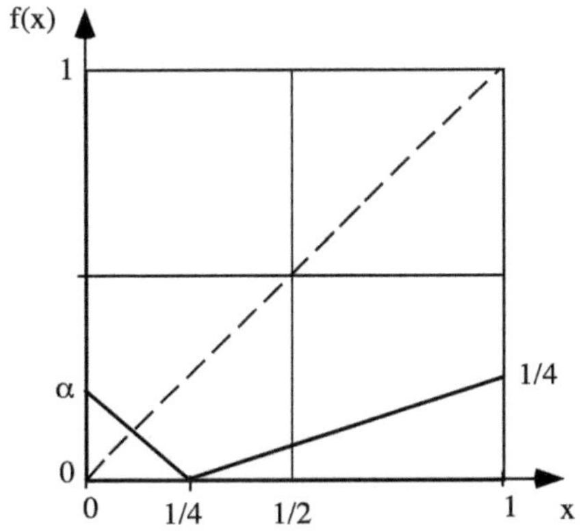

Consider the dynamical system $x_{t+1} = f(x_t)$.
(a) At what value on α does the system become chaotic (starting with α being small)? Characterise the dynamics for α below the critical value for chaos (stable/unstable fixed point and/or periodic). Is there any other value on α that corresponds to a drastic change in the dynamics in the non-chaotic regime?
(b) Suppose that $\alpha = 1$. Determine the invariant measure that characterises the chaotic behaviour, and calculate the Lyapunov exponent λ. Calculate also the measure entropy from the finite state automaton describing the symbolic dynamics (for a generating partition).

(c) If you know that the system is in the region $x < 1/4$ at time t, how much information do you get if you observe the system in the same region again at time $t + 3$ (again assuming $\alpha = 1$)?

8.4

Let a map $f(x)$ be defined by the figure below, where $1/2 < \alpha < 1$.

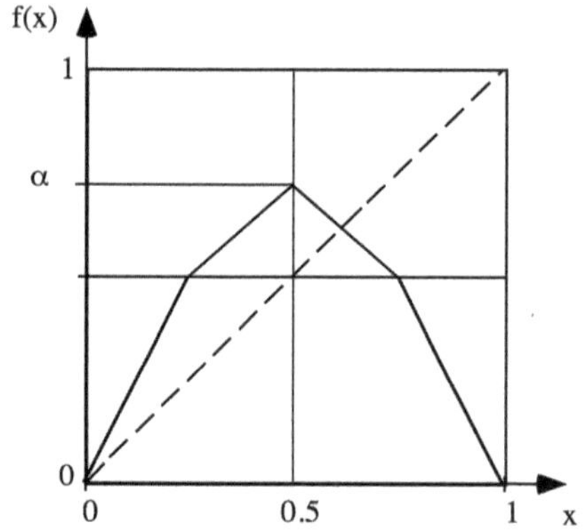

Consider the dynamical system $x_{t+1} = f(x_t)$.
(a) At what value on α does the system become chaotic?
(b) Suppose that $\alpha = 7/8$. Determine the invariant measure that characterises the chaotic behaviour, and calculate the Lyapunov exponent λ. Calculate also the measure entropy from the finite state automaton describing the symbolic dynamics (for a generating partition).
(c) If you know that the system is in the region $x < 1/4$ at time t, how much information do you get if you observe the system in the same region again at time $t + 3$?

8.5

Let a map $f(x)$ be defined by the figure below (so that $f(0) = \alpha$, $f(1/4) = 1/2$, $f(3/4) = 1$, $f(1) = 0$, with $0 \leq \alpha \leq 1/2$.

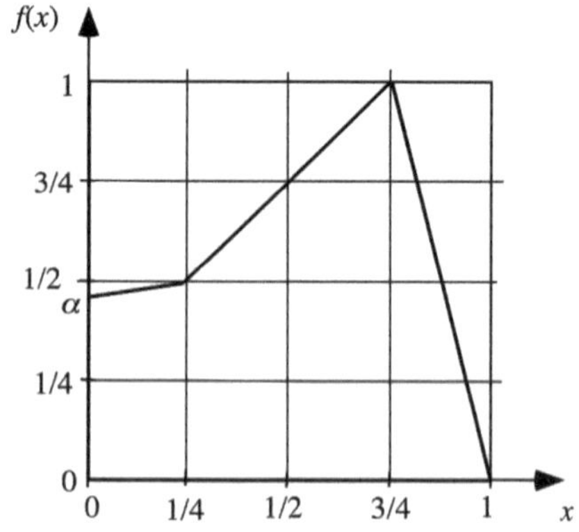

Consider the dynamical system $x_{t+1} = f(x_t)$.
(a) At which value of $\alpha < 1/2$ does the system become chaotic when decreasing from $1/2$?
(b) What is the behaviour for α close to $1/2$ (above the critical value derived above). Describe qualitatively only.
Assume from now on that $\alpha = 0$.
(c) Find the invariant measure μ that characterises the chaotic behaviour, and determine the corresponding Lyapunov exponent λ by using that measure. Find a partition that has a symbolic dynamics with a measure entropy s_μ that equals the Lyapunov exponent (as one should expect).
(d) Suppose that we at a certain time t observe the system in the region given by $x > 3/4$. If we find the system in this region again three time steps later (at $t + 3$), how much information do we gain by this observation?

8.6
Suppose that for a continuous map $f(x)$, $0 \leq x \leq 1$, holds $f^2(0) = 1$, $f^3(0) = 0$, and that the function is linearly increasing in the interval $0 \leq x \leq f(0)$, and linearly decreasing in the interval $f(0) \leq x \leq 1$.
Consider the dynamical system $x_{t+1} = f(x_t)$.
(a) Determine at what value of $f(0)$ the Lyapunov exponent λ has its maximum. What is the value λ?
(b) If an observer knows that the system at time t is found in the interval $0 \leq x \leq f(0)$, how much information does the observer gain when she also learns that the same holds at time $t + 4$?

8.7
Let a map $f(x)$ be defined by the figure below, where $f(1) = \alpha$ and $f(2/5) = 0$.

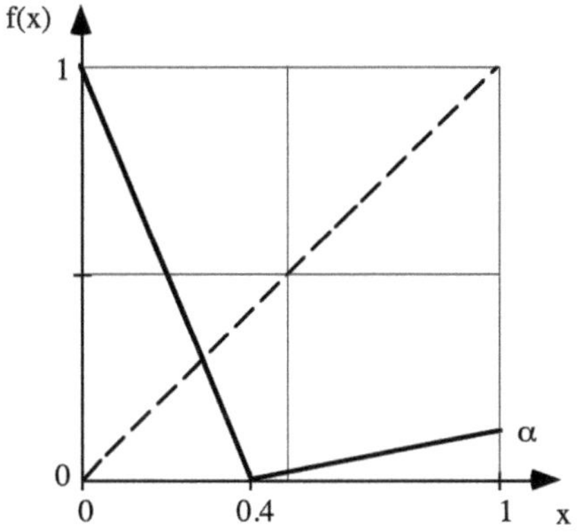

Consider the dynamical system $x_{t+1} = f(x_t)$.

(a) Describe the behaviour when α is small? At what value of α does the system become chaotic?
(b) Suppose that $\alpha = 2/7$. Calculate the Lyapunov exponent λ. Find a generating partition of the interval [0, 1]. What is the measure entropy s_μ?
(c) What is the Lyapunov exponent if $\alpha = 2/5$?